P9-DWA-283

PLACE AND BELONGING IN AMERICA

PLACE AND BELONGING IN AMERICA

David Jacobson

THE JOHNS HOPKINS UNIVERSITY PRESS
BALTIMORE AND LONDON

Printed in the United States of America on acid-free paper
9 8 7 6 5 4 3 2 1

The Johns Hopkins University Press
2715 North Charles Street
Baltimore, Maryland 21218-4363
www.press.jhu.edu

Library of Congress Cataloging-in-Publication Data

Jacobson, David, 1959–
Place and belonging in America / David Jacobson.
p. cm.
Includes bibliographical references and index.
ISBN 0-8018-6779-7 (hard : alk. paper)
1. Human geography—United States. 2. Ethnology—United States.
3. Values—United States. 4. Pluralism (Social sciences)—United
States—History. I. Title.
GF 503 .J3 2001
304.2′3′0973—dc21 2001000540

A catalog record for this book is available from the British Library.

To Adam, Maya, and Noam

CONTENTS

ACKNOWLEDGMENTS

WRITING THIS BOOK has been a great joy. Much of that joy has been due to the participation of the wonderful people whom I have talked to and exchanged ideas with and who have been a warm presence along the way. Surely one of the privileges of writing a book is being able, through the acknowledgments, to lighten, if not remove, the moral accumulation of the writer's debts.

So many have helped me with this book, a book that had been germinating in my mind long before I fully articulated it, from many conversations over many years. I am deeply grateful to Steven Grosby for his extensive and ongoing comments on this work. I benefited greatly from the remarkable scope and depth of his scholarship. I do not know how many times I talked with Jamie Goodwin-White about themes related to this book (there were countless such times), but it does not really matter; those conversations make up one golden thread that is woven into the fabric of the book itself. Michael Kearney sent pages and pages of single-spaced comments that were themselves a feast for the mind. My friends at the Copenhagen Peace Research Institute, where I spent a year as a visiting fellow just before the turn of the twenty-first century, were wonderful and stimulating colleagues. The city itself was such an inspiration. In Copenhagen, Dietrich Jung provided numerous ideas and thoughts from which I profited greatly, and Jesper Sigurdsson and Ole Wæver were sources of probing thoughts and fruitful suggestions. Henry Tom was a most supportive editor, pushing me forward at the right times. My students at Arizona State

University were always patiently ready to listen and to put forward their observations.

Let me mention all too briefly my deep gratitude for their thoughts and suggestions at different times to George Thomas, Karen Miller-Loessi, Saskia Sassen, John Meyer, Ari Zolberg, Suzanne Keller, Thomas Berger, Didier Bigo, Carolyn Forbes, Eric Mumford, Kevin McHugh, Haaken Wiberg, Steve Mackinnon, Joel Gereboff, Steve Cornell, Subhrajit Guhathakurta, Edward Linenthal, John Torpey, Galya Benarieh Ruffer, Jim Holliefield, Joe Rhea, Zhenchao Qian, Victor Agadjanian, Galya Lahav, Yosef Lapid, Martin Heisler, Mathias Albert, Lothar Brock, Rey Koslowski, Edward Lebow, Vivien Spiegelman, Gray Sweeney, and Paul Wapner. I am grateful to Ellen Sealey, Sara Stebbins, and Stewart Fritts, of the National Park Service, for a lengthy interview by the Grand Canyon. I am also indebted to Grace Buonocore for her careful copyediting work, and to Zeynep Kilic and Paula Walsh for their extensive research assistance.

I thank different hosts for their kind invitations to present different parts of this work as it progressed at workshops and conferences at the University of Geneva; University of Chicago; New School of Social Research; European University Institute; Cycladic Academy for Europe, in Santorini, Greece; University of California at Irvine; Hebrew University of Jerusalem; New Mexico State University; Boston University; University of California at Los Angeles; and the Centre d'Études et de Recherches Internationales (CERI) in Paris.

For other friends this book will be mostly new. But their very presence and company have always been a font of great encouragement and warmth. I want to thank in particular four close friends of many years: Paul Lunney, Gina Dominique, Lori Schneider, and Kirsten Evans.

To Sharon, what can I say adequately: to her belongs my gratitude for her support and endless patience. I dedicate this book to our children, whom we love dearly. They bring us wonder and happiness.

PLACE AND BELONGING IN AMERICA

Introduction: Terra Firma

IN THE FIRST SEVEN DAYS described in Genesis we have a poetic distilling of the human condition. Space and time, heaven and earth, right and wrong. As God created a material universe, he created a moral one as well; space is not merely physical but moral. In the beginning, when the earth was without form, we had darkness, an existential as well as a terrestrial void. Indeed, from the original Hebrew the term for *without form* could equally be translated as *chaos.* But then God began creating forms, distinctions, and light, and with them the possibility of knowledge of truths and of God himself. He created night and day, dry land and seas, days and years. Physical form and moral place are inherently related to each other. In Genesis, in other words, place is moral as much as it is physical; and humankind's association with geographic place, and the forms that association takes, tell us much about the character and shape of different social identities.

The relationship of physical space to humanity's "place" in the cosmos is not a purely religious concern; Genesis captures a fundamental sociological truth about the interplay of geographies—moral, social, and physical. It is at the intersection of these geographies that this book is situated. How does a community come to be identified with a certain territory, and how does a territory "define" a people? Why are a people so intricately associated, in the morally felt sense of *rightfully* belonging, with a certain place, such as the Americans with the United States, the Jews with Israel, or the French with France? And, just as important, what happens, as is now happening, when peoples and their lands become uncoupled, that is to say, when communities increasingly live outside their imputed homelands, in diasporas? What does such an uncoupling bode for the boundaries of community and the shape of politics?

People are moving across borders voluntarily to a degree unprecedented in recent history, and foreign populations are growing in areas such as Western Europe and North America. Multiple loyalties are becoming increasingly common in the form of dual citizenships of many immigrants and travelers. This reflects in part new legal sanctions allowing multiple citizenships in countries where singular loyalties were once the law. The sense of singular national belonging once explicit in citizenship is increasingly a shadow of what it once was. The phenomenon of people no longer feeling a special moral association with the land in which they live is becoming fairly common—their house is no longer their home.[1]

This book asks two questions, focusing on the United States. First, how did the American people come to develop a moral association with this land, such that their very experience of nationhood was rooted in, and their republican virtues depended upon, that land? Second, what is happening now as the exclusivity of that moral linkage between people and land becomes ever more attenuated? How are the political process and forms of community recast in this context?

The book is divided into two parts, dealing with these respective questions. The first part considers particular times or moments of American history: the Puritans and the idea of America; the early Republic and agrarian virtue; the nationalization of place in the Civil War and beyond; and the post–Civil War monuments and conservation of national parks. The second part concerns the contemporary United States. We consider in this part the more multifaceted relationships to the land, in which different cultures are sewn into a shared landscape—which in turn has profound implications for the form of civic political life. The kaleidoscope of cultural expression is examined as it is portrayed in the organization of public space—be it urban or suburban spaces, public squares, monuments, memorials, or parks. We also consider the changing patterns of immigrant incorporation, reflecting a more polymorphous human landscape. Throughout this book, the political dimension is central: the emergence of republican politics that are fundamentally rooted in a bounded territoriality; and, more recently, how politics have been transformed as the exclusivity of people and land wanes—a transformation that has accentuated the role of the courts and judicial politics.

The Sense of Place

It is difficult, for those in Western countries at least, to understand the multilayered nature of place: its geographic, social, moral, and economic qualities; its promise of order, functional and metaphysical ("to know one's place"); and its designation of not only a location but also a state of being. One's place is one's home. Home is a refuge and a point of departure from, and orientation to, the world. Home is existentially where one is secure. In the words of an anonymous poet:

Jerusalem, my happy home
When shall I come to thee?
When shall my sorrows have an end,
Thy joys when shall I see?

Place, home, soil. These allusions are often biblical, with the world-to-come portrayed as the "long home" following the journey of life: "Then shall the dust return to the earth as it was: and the spirit shall return unto God who gave it" (Eccles. 12:1). In contrast to being in one's place, travel is portrayed in terms of its root meaning, a travail. Such wrote one poet, "Half to forget the wandering and the pain . . . and dream and dream that I am home again!"[2] Place has a distinctly moral element, containing as it does notions of belonging, of one's rightful place in the world, locating individuals and peoples geographically and historically and orienting them in the cosmos. Place suggests that the cosmos is ordered, not an infinite stream of space and time. Place links the metaphysical sense of belonging with the physical present; the act of orienting spatially and temporally is fundamental to the human condition.[3]

The difficulty in understanding the different nuances of place is due to the way the idea of the nation-state has so blended the geographic, communal, and political dimensions of belonging as to make them indistinguishable. The multilayered meanings of place seem, or have seemed, so direct and uncomplicated in the modern world. Nation-states from their beginnings in seventeenth-century Europe knitted the territorial, national, and political strands of collective identity into whole cloth. The Dutch republic from its independence in 1648 represented territorial integrity, a new republican politics, and a newly expressed sense of religiously based nationhood at one and the same time. The American Revolution in 1776 demarcated a territory, established a new people, and declared a new form of government in, so to speak, a single act. The French Revolution did the same in 1789. The political systems in each case were distinctly different, though each was republican in the broad sense. Nevertheless, nation and state, the people and the land, came to be conceptualized as an integral whole, singular in their unity.

The marriage of nation, state, and territory has often led political analysts to reduce the state to a single dimension, often the territorial. But as peoples' moral associations with geography change (for example, living as diasporic peoples, or imagining themselves as such, with their symbolic centers lying elsewhere), the very nature of politics, and of community, changes.

This juxtaposition of community, territory, and polity is a relatively new phenomenon in history and until the nineteenth century was the exception rather than the rule.

For American Indians, for example, political organization was traditionally only one aspect of group identity. Political power was commonly located at the level of the village, band, or kinship group. But collective identity went well beyond these social units. Tribes or nations were much wider-ranging and more comprehensive groups than the village, and they shared common identities, symbols, cultural practices, and networks, which girded a common identity.[4] As Ernest Wallace and Hoebl Adamson note of the Comanches, who were made up of autonomous bands and families, "'Tribe' when applied to the Comanche is a word of sociological but not political significance. The Comanches had a strong consciousness of kind. A Comanche, whatever his band, was a Comanche . . . By dress, by speech, by thoughts and actions the Comanche held a common bond of identity . . . In this sense the tribe had meaning. The tribe consisted of a people who had a common way of life. But that way of life did not include political institutions or social mechanisms by which they could act as a tribal unit."[5] Thus the sense of collective identity was generally much more comprehensive than the political and organizational basis of such Native American groups. Territory, similarly, often had no matching political organization. Some groups had claims to exclusive tribal territory, whereas others did not or articulated them only under the pressure of the territorial expansion of whites. Lands could also have varying and graduated meanings, especially for more nomadic peoples, involving notions of sacred space, economic space wherein resources were obtained, and "common space," whereby space may be shared with other groups on a competitive basis.[6]

Other cultures and civilizations also varied dramatically from Western concepts of boundedness and of juxtaposing community, territory, and polity. Imperial Russia, for example, had no concept of "natural boundaries." Instead, Russia conflated images of nation and empire. Christianity became equated with Russia itself. Russian expansion into and colonization of the hinterlands began apace, under Tsar Ivan III (1462–1505). Migration and expansion came to characterize Russian history. Expansion was a national enterprise; it lasted centuries and involved constant movement to new lands. Similarly, Muscovy engaged in centuries of unrelenting military conflict. Defense became the raison d'être of the all-powerful state.[7] Pluralistic Western concepts such as national self-determination were foreign to the Russians. Likewise, the concept that a doctrine, such as national rights, was a universal principle was equally alien. Moscow, prior to Peter the Great's reign (1682–1725), consequently avoided entering into diplomatic relations with foreign powers. Russian "nationhood" was completely at odds with the Western concept. The Russian nation was viewed as being coextensive with the Russian empire, an empire that was, in the words of Richard Pipes, "sui generis, unique and unrelated to any other, part of no state system or international community, the only guardians of the true orthodoxy."[8]

Confucian China was even more radically different from the nation-state. For the Chinese, the world was Sinocentric, "all-under-heaven," and was presided over by the emperor, the Son of Heaven. China was literally the Middle Kingdom, at the center of a spatial hierarchy of relations with the external world. The emperor was not only a temporal ruler but a cosmic figure as well. He was a mediator between heaven and earth, a mediator of cosmic harmony. The emperor occupied a cardinal point in a universal continuum. The Middle Kingdom was an "empire without neighbors." Civilization and barbarism lay conceptually on the same continuum—what was not civilized was barbarian. *Chinese* civilization, as such, did not exist; just "civilization."[9] The Chinese incorporated the rest of the world into their own. The idea of a corporate "people," or nation, was alien to Confucian China. Being "Chinese" as such was of little consequence. Non-Chinese conquerors could be absorbed into the Confucian system *and* be able to rule. Some did, such as the Manchus after 1644.[10] The Middle Kingdom was the cultural, as well as geographic, center of the world. Similarly, the idea of fixed frontiers and the definitive and sudden transition of the modern state border were absent.

Relations with surrounding areas, and with barbarians in general, proceeded with assumptions of Sinocentrism and Chinese superiority. The concept of foreign policy did not exist. Domestic and external relations were graduations on the same continuum. External relations reflected social and political principles indigenous to Chinese society—principles of inequality and hierarchy. The Q'ing dynasty (1644–1912), the last before imperial China collapsed, perfected the tribute system for conducting trade and external relations. The tribute system called for an elaborate ritual that symbolized the inferiority of external envoys and traders. Russian and British envoys who refused to kowtow were labeled rebels. Order in the Middle Kingdom was dependent on the barbarians fulfilling their role in the scheme of things.[11]

Europeans, in contrast, believed in a universal truth, but civilization was geographically and cognitively bounded. For the Europeans the barbarians were to be subjugated and converted, not pay tribute. Furthermore, China displayed a passivity, inherent in the makeup of the Middle Kingdom, showing little interest in discovering, conquering, or extending its influence into barbarian lands. The Sinologist Joseph Levenson contrasts European dynamism with Chinese insularity:

> In European history we find the Christian transcendental sense of divinity and evolutionary sense of history, then the modern secular messianisms with their visions . . . of progress in time . . . culminating in progress in space, outwards from Europe. In Chinese history we find Confucius, for whom "Heaven does not speak" but rather reflects a cosmic harmony as a model of society . . . Against the transcendental and the evolutionary, we must set Confucian immanence and orientation to the past. Nothing repelled the normative Confucian more then messianic goals and eschatological structures . . . The meaning of history was not the end-stage of culture but in sage-antiquity.[12]

The Chinese were their own Jerusalem. No tension of a transcendent promise that had to be attained was present.[13]

In the "transcendental" West, in contrast, in the wake of the Reformation a more dynamic notion emerges of corporate peoples, imbued with the grace of God (and, later, with secular nationalist faiths), who are driving towards a millennial eschaton, an End of History, redeeming an essentially corrupt world by establishing the Word—the law—of God. By politically and territorially establishing themselves, Dutch Calvinists, the Scottish Presbyterians, and the Puritans, among others, sacralized their lands through the assertion of the power of Yahweh and his law. Just as the people were bounded through their special Covenant with God, so the land they settled or occupied was delimited: sacrality was not unlimited, and jurisdictions had to be established. The law—the law of God—and by extension, the polity, were centrally involved in this project. Politics and control of the earthly kingdom became a calling. In this sense, these Protestant groups replicated the experience of the ancient Israelites. As the sociologist Steven Grosby writes of the Israelites, "The entire bounded territory was believed to be the possession of Yahweh; thus, the Promised Land as a whole was believed to be sacred."[14]

These Protestant notions were in stark contrast with the medieval monasticism of the Catholic Church, whereby the world, corrupt and unredeemable, was to be "escaped." Territoriality—in the sense of a bounded, regulated territory that defined a people or religion—was inconceivable. The church was, in principle, universal and unbounded, and transients could pass through territories relatively unimpeded.[15] Travel was not bimodal—from one territory to another—but cyclical, as in the pilgrimage: one did not "immigrate" but went on pilgrimages (or even crusades), to return, it was hoped, with a cleansed soul.

Later secular nationalisms reproduce this belief of the people and the land tied together in an organic and sacred unity. The nation replaces or supplements God as the transcendent truth; the individual's life takes on greater meaning and resonance in the historical epic of an immortal nation driving towards, if not the millennium, a golden future. The citizen replaces the Calvinist idea of the politically active saint.[16] As religious faith declines, war memorials and national monuments come to symbolize and express nation and country. The term *domestic* is extended to cover vast countries (as in domestic policy), and the sense of kinship with the land is so powerful that the term *homeland* becomes common parlance. Indeed, blood and soil become interchangeable in nationalist expression. As Rupert Brooke, who died in 1915, wrote:

> If I should die, think only this of me:
> That there's some corner of a foreign field
> That is for ever England. There shall be
> In that rich earth a richer dust concealed;
> A dust whom England bore, shaped, made aware,
> Gave, once, her flowers to love, her ways to roam,
> A body of England's, breathing English air,
> Washed by the rivers, blest by suns of home.

The land and the people become one. Law maintained its sacred quality, carefully demarcating the national jurisdiction; but the law now reflected not so much a deity as the "will of the people." The religious element, however, remained central in some nationalisms, such as in the United States or among the Afrikaners in South Africa.

Out of Place

The sentiment of *exile* is the corollary of being rooted in native soil: The ache, aimlessness, and despair of exile, to be outside one's country, to be banished to the wilderness, is a deep and lengthy literary theme in human history. The image of the Jews, exiled in the sixth century BCE to Babylonia, weeping in anguish on the riverbanks and yearning for their homeland, has had a powerful resonance in Jewish and Christian history. The theme of exile is, indeed, central to the biblical story, as is the tie of a people to the soil. The prophet Isaiah (27:6) reassured the exiled people of Israel, "In days to come Jacob shall take root, Israel shall blossom and bud; and all the world will be filled with fruit." The conception of *rootedness* is almost literal—a people rooted in their land.

The sense of exile outside one's home, of uprootedness, goes well beyond the Jewish and Christian traditions, which informed the nation-state, even though the tie with the soil may be configured in ways that are distinctly different from nation-states. A Navajo Indian chief, Barboncito, told the Civil War general William Tecumseh Sherman what banishment to lands outside their native soil of Navajoland, in what is today northeastern Arizona, northwestern New Mexico, and a fragment of Utah, had done to his people:

> Our Grandfathers had no idea of living in any other country except our own . . . When the Navajos were first created four mountains and four rivers were pointed out to us, inside which we should live, that was to be our country and was given to us by the first woman of the Navajo tribe . . . Our God . . . gave us this piece of land and created it specially for us and gave us the whitest of corn and the best of horses and sheep . . . This ground [in the reservation far from the Navajo lands] we were brought on, it is not productive . . . It is true we put seed in the ground but it would not grow two feet high, the reason I cannot tell, only I think *this ground was never intended for us.*

In a poignant plea, Barboncito says to Sherman, "I want to go and see my own country. If we are taken back to our own country, we will call you our mother and father, if you should only tie a goat there we would all live off it, all of the same opinion."

Sherman responded by saying, "You are right, the world is big enough for all the people it contains and all should live at peace with their neighbors," and added, "All people love the country where they were born and raised." Barboncito noted at the end of Sherman's response: "I hope to God you will not ask me to go to any other country except my own." After it was clear that Sherman would agree to the Navajos' return, Barboncito said, "After we get back to our country it will brighten up again and the Navajos will be as happy as the land, black clouds will rise and there will be plenty of rain. Corn will grow in abundance and everything [will] look happy."[17]

Republican Politics

The moral connection of a nation-state to a bounded territory, however, was not only religious or nationalist but also inherent in the *political* culture that began to emerge in seventeenth-century Europe. This culture of civic or republican politics blossomed with the American and French revolutions.[18] There is extensive and ongoing discussion on civic politics or civil society—particularly now when it is perceived to be in decline—but one aspect of civil society is usually ignored or at best implicit, namely, its territorial dimension. Civic societies are, at least as they are presented, made up not only of active citizens but also of persons bound together on the basis of voluntary associations, such as churches, political parties, unions, sororities, and parent-teacher associations. Such associations are also bound together by relatively abstract ideas and beliefs of a political, social, or religious character.

Contrasting civil societies with the other major form of social organization, kinship, brings the nature of the former into sharp relief. Kinship is the principle that has governed societies from hunting-gathering bands to feudal Europe (albeit in the latter case with certain qualifications, noted below). When the principle of kinship governs political rule as well as marriage and inheritance, power is expressed through personal relations. This concerns not only the control of power through family descent but also the execution of power: it depends on the direct personal ties of those in leadership—a prince, chief, village elder, or patriarch, for example—with those directly under the leader, such as feudal vassals, or with those with family and clan affiliations. Thus political power tends to be relatively limited and rather diffuse. Associations are on the basis of blood and marriage and are not voluntary, nor are they based on abstract ideas or ideologies.

In the civic polity, territory becomes the counterpart to kinship. Translocal or national loyalties could not be built on the principle of kinship. Bounded territoriality is one critical thread in sewing together potentially disparate persons into a single entity, a border that is also a social and political marker indicating commonality. Politics becomes nationalized and ultimately public, as well, and power comes to inhere in offices, not the officers. Kingdoms based on kinship could, in the past, migrate, and some did, such as the Zulus in Africa and the West Goths in Europe.[19] When political identity was rooted in the land, and not in kinship ties, the state was relatively fixed (relatively, because such states did seek to aggrandize themselves). People then migrated, and immigration involved shifts in loyalties and affiliations in a way that before it did not.

At the time when territory was generally held to be of mundane significance, only certain holy sites were considered sacred, points where deities, or the Deity, could be accessed. Thus in the feudal period, and in the Catholic tradition, pilgrimages were common. The Protestant reformers, on the other hand, condemned pilgrimage.[20] With the territorialization of faith and the rise of religious (Protestant) nationalism, the land as a whole became sacred. One does not have to be at a particular site to experience the sacred presence. In contrast to the principle of kinship, territoriality becomes the marker of "bounded, spatial communion." What Steven Grosby wrote of ancient Israel is equally applicable to emerging nation-states in post-Reformation Europe: "The land having been possessed by a people, is part of a people. The land is a spatial focus of the memories of the people and of the present and future existence of the people . . . Through the act of possession of an area of land, and decisively through the inhabitants' reflection on that possession, an area of land becomes *their* land, it becomes significant; it becomes a territory."[21]

The radical distinctiveness of kinship and civic-territorial principles is palpable today, especially among formerly colonized peoples who are forced to try to adapt, often in a historical moment. Many parts of Africa have been rent by this process. But even in the domestic domain of the United States this juxtaposition of principles is evident. As in most states in Africa, the political frameworks (and the borders of reservations) imposed upon American Indians are often incongruent with underlying communal identities. The sociologist Stephen Cornell writes how structures of "authority and decision-making, once imbedded in the fabric of aboriginal societies, [are] now attached, as it were, from the outside, institutionally separate from the structures of kinship and custom and modes of thought which 'governed' Indian peoples." The Oglala in South Dakota, for example, have little faith in representative forms of government and prefer local, community leaders. "Full-blood" Cherokees in Oklahoma pay little attention to the official tribal government and do not participate in it.[22]

Caveat Emptor

An important caveat to this connection between territoriality and republicanism should, however, be noted: recently, historians and sociologists have been vigorously debating whether nations and national identity are relatively recent phenomena or whether at least some of today's nations are premodern and even perennial. Some "perennialists" argue that the phenomenon of nations is found everywhere in history, though particular nations may come and go. Others suggest that present-day nations extend back centuries or even millennia. Those who stress continuity of nations suggest that the conventional view of the Middle Ages as purely kinship-based is incorrect.[23]

The medievalist Susan Reynolds, for example, suggests that after the year 900, kingdoms with fairly well defined territories and ruled by a monarch were associated with peoples who had some sense of collective identity. This sense of such an identity was based on a perception of shared descent and custom. Society was not solely "vertical"—the serried and hierarchical ranks of lords defined down to the peasant (with all ranks determined by and large through kinship); it also had horizontal, communal bonds. Government was exercised in layers, with rights and obligations at every layer, and the leading men of each community were assumed to represent it (such that a nobleman could represent his community, not just his rank). This was the case even if noblemen were not elected to do so in the modern, democratic sense. Medieval society thus valued a combination of both lordship and community.[24]

It is, then, important to qualify the pointed *analytical* distinction between kinship and civic-territorial principles in terms of the more measured *historical* context. Because the movement from kinship to civic-territorial principles has been historically more graduated, the distinction between traditional (or premodern) and modern has to be considered and moderated in terms of the specific historical period concerned. Thus the emergence of civic-territorial principles may have been more or less rapid or may have involved what are viewed in retrospect as precursors of what came afterward. Hence what Reynolds describes of medieval society, with its territorially contiguous communities, may indeed be an important corrective to conventional medievalist understanding. However, clearly her description of medieval society is still a good distance from the civic-territorial principles that came afterward. As Reynolds herself notes, for people of low rank the local community was most real; indeed, the "state" as represented by the king was distant and opaque for the majority of the population, whereas their local community, family, and church were tangible and definite. In republican states that would come later, the state is more critical for the individual's livelihood and identity than perhaps even his or her family, and certainly more important than the church or local community.[25] The idea of the state, representing a *sovereign* people, was absent in the Middle Ages; similarly, it is only with the modern state, if contrasted with medieval society, that we see the direct and unmediated legal relationship between the individual and the state.

Even the careful demarcation of territorial boundaries becomes more defined, more rationalized, and of concern (and technically possible) after the Middle Ages. The geographer Robert Sack observes that in medieval life the conception of place, space, and distance was tied to the palpable and literal experience of everyday life. "Events and experiences" defined mental maps, not the "location in abstract space, defined places and distances" as in republican nation-states. Sack notes that when distant places were mapped together, medieval maps tended to picture a religious cosmography. The world was usually divided up into three regions—Africa, Asia, and Europe—with Jerusalem often at the center. The spiritual center, not some sense of the physical universe, defined these maps.[26]

Civic Evolution

In the United States at the turn of the twenty-first century, participation in various civic and voluntary groups appears to be declining. Studies suggest that even simple social engagement is down. Political participation, particularly voting, is historically low. In both the general media and academia, this decline in civic membership and voluntarism and its causes have been extensively discussed. Television, for example, has been blamed as a major force in this decline.[27]

Curiously, this debate has been almost completely detached from discussion on changing conditions associated with foreign populations, growing transnational associations, and the changing forms of civic and political life. Over a period of thirty years in the United States the value attached to citizenship (in terms of the affiliations it demands) has been evolving in new directions.[28] Dual citizenship is becoming more common. As the traditional conception of citizenship as a singular loyalty fades (for these processes implicate natives as well as foreign-born residents), so the moral tie between land and people is attenuated or, at least, conceptualized in less exclusive, felt ways. In other words, the apparent decline in participation in civic organizations and the weakening of territorial identities may well be intricately tied to each other.[29] Social and communal boundaries are becoming quite distinct from territorial boundaries. Political models associated with the nation-state—of the state embodying the "general will" and the will of the people—are becoming less able to accommodate recent social developments. Related to this, and an issue that is addressed further below, the growing density of judicially mediated rights, centered on the individual as opposed to the collective body politic or nation, makes the process of voting and legislative give-and-take less critical to the individual's legal and social privileges and prerogatives. Thus the decline of voting occurred exactly when the expansion of "judicial rights" took place starting in the 1960s.

Moral Territories

Borders and boundaries designate moral proximity and moral distance, inclusion and exclusion.[30] Boundaries, be they geographic or nongeographic, are forms of "markers"—visible signs, as Erving Goffman described them, of a "territory" of some kind. Markers can delineate physical borders, but clothes, architectural designs, an umbrella on a beach, dietary laws such as kashrut or vegetarianism, and so on are also markers. Markers can be fixed (like state boundaries), transient (space on a beach), or transportable (like clothes).[31] "Space" and "social distance" become the elements that, in a sense, define social and political forms.[32] Marriage, friendship, comradeship, kinship, conflict, work, play, notions of private and public, and the like indicate varying forms of association and social spacing. The social order is implicitly about spacing, about maintaining the "proper" patterns of participation. This characteristic of social order is apparent in the most pedestrian of human circumstances:

> A condition of order at the junction of crowded city thoroughfares implies primarily an absence of collisions between men or vehicles that interfere with one another . . . [Order] does not exist when persons are constantly colliding with one another. But when all who meet or overtake one another in crowded ways take the time and pains needed to avoid collision, the throng is orderly. Now, at the bottom of the notion of social order lies the same idea. The members of an orderly community do not go out of their way to aggress upon one another. Moreover, whenever their pursuits interfere, they make the adjustments necessary to escape collision and make them according to some conventional rule.[33]

What is striking about social distance on the most personal level is that it is intricately related to the shape and form of society as a whole. For example, a society that prides itself on the rights of privacy and the individual will conversely limit the public reach of government. As the German sociologist Georg Simmel wrote, "[A] sort of circle . . . surrounds man [and] is filled out by his affairs and characteristics. To penetrate this circle by taking notice, constitutes a violation of his personality . . . The question where this boundary lies [around the individual] cannot be answered in terms of a simple principle; it leads into the finest ramifications of societal formation."[34] Thus, moral distancing and closeness are the very warp and woof of the social fabric: "Concord, harmony, coefficacy, which are unquestionably held to be socializing forces, must nevertheless be interspersed with distance, competition, repulsion, in order to yield the actual configuration of society."[35]

Conversely, in Maoist China, for example, the self was subsumed under the collective interest. "Comradeship" became the mode of personal relations in place of friendship.[36] Comradeship implies a universal morality wherein all are equal. Friendships imply the presence of a "particularistic" universe closed to outsiders and thus purportedly threaten the collective communal interest. Instead, a norm of universal civility and helpfulness is to be encouraged.[37]

Thus there are "moral territories," almost by definition, in any social order. But these moral cartographies or maps do not have to be coextensive with physical territories, as in the nation-state. As Stanford Lyman writes, "Interactional territories may be spatial but [may not be] necessarily cartographed on a geopolitical map." In a world of massive migrant and refugee flows, global productive processes, transnational ethnicities, and the Internet, this point becomes abundantly clear.[38]

In the nation-state, we noted, territory has served as a social space (in the form of the national community) as well as a geographic demarcation; or, to put it a another way, territory has served to define the nation, it history and memory, its place and location, in the moral as well as in the literal sense. With the erosion of territory as the preeminent marker of political and national community, we need to ask ourselves what nongeographic forms of "social space," of community, are replacing or supplementing geographic markers. Moreover, what is the meaning of "politics" in this context, given the inherently geographic character of the republican, civic polity associated with the nation-state?

New Social Markings

The political theorist Michael Walzer once warned that if national borders were not maintained as markers of a national community, internal distinctions and demarcations would become more pronounced. The admission or exclusion of immigrants, he suggested, was at the very core of "communal independence." The sense of a common life, of a historically stable, ongoing association, and of a community with a special and mutual commitment could not be maintained without such markers. If these national markers weakened, internal boundaries would come to the fore to the point of insular "fortresses" emerging.[39] Walzer has proved to be at least partly correct, though the darker side of this picture appears true for only, say, parts of Los Angeles and some other urban areas.

As national borders become less significant as communal markers, domestic boundaries and markers (not necessarily of a geographic kind) become more pronounced. The idiom of multiculturalism may be in contention, but there is increasing recognition that the melting-pot ideal cannot be resurrected.[40] In other words, communal distinctions have been increasingly domesticated and politicized (and even internationalized; these communal identities often cut across national borders). Patterns of association and dissociation, and of moral linkages and breaks, are thus shifting in ways that cannot be easily fitted onto a political map. This development reflects a world where diasporic communities are increasingly common and where once international divisions of north and south, of developed and developing countries, have been domesticated.[41]

Thus something akin to, yet obviously different from, the medieval skein of borders and boundaries may be emerging, with interlocking and overlapping social and political entities increasingly the norm today. In this regard, international relations cannot be viewed as something distinctly political (or, at best, political and economic), as diplomats have traditionally viewed their domain, free of cross-national communal and social "complications." Like the medieval prince but in a different way, the state finds today that social and political issues—from environmental matters to human rights and immigration—cannot be easily untangled. Domestic, even local, initiatives on, for example, illegal aliens in California produce riots in Mexico City. Thus, as social boundaries come to cut across geographic borders, so social and communal ties are more and more implicated in international politics.

The emergence of territoriality is translocal; it promotes national or regional identities in place of or supplementing local, village, feudal, or kinship identities. Conversely, as territoriality becomes less significant in marking and defining (national) community, so local and transnational communal identities are asserted (or reasserted).

New Political Landscapes

How are these new social and communal groupings marked, if not in geographic terms? They can be marked through language, ethnic markers (clothes, speech, associations), body language, codes (on, for example, notions of sexual harassment), and disciplinary distinctions in universities and church and organizational affiliations—what Goffman aptly called the "territories of the self."[42] What is striking about this form of boundedness is that it is fluid, portable—a constantly shifting, yet defined, road map. It can be group-based, or it can be individual; in a multicultural world, it is predicated on the discourse of "rights," and this is important for its political configuration. Rights are about personal distance—from government but also from one another. Rights accentuate ethnic, cultural, and gender distinctions. In contrast, collective notions of nationhood assume a commonality and tend to be expressed in legislative acts. And rights ultimately—as in human rights—are portable and are not predicated, as is nationhood, on territoriality.

What shape, then, does politics take?

The "judicialization" of politics—the growing role of the judiciary and other administrative mechanisms in political life—in the United States (and also in Western Europe and other parts of the world) is inherently tied to the growing stress on "social space" and personal or group distance (as expressed in, for example, multiculturalism) on the one hand and the "deterritorialization" of political and social relations on the other. Judicialization is the growing ability and readiness of courts to review legislative acts and to void them when they have been deemed to violate constitutional norms. In this context courts take a more proactive role in setting standards for the adjudication of different issues.[43] Even in the United States, where the courts have traditionally wielded formidable sway over the federal and state legislatures, the growing importance of rights as law and rhetoric has added to the courts' significance. The judiciary places bounds on the actions of government vis-à-vis the individual; it delineates personal distance, between state and individual or groups and between individuals and groups themselves. The idiom of judicial (or constitutional) politics is one of "rights." But judicial politics goes beyond the courts. It is reflected in the growing role of government administrative bodies, in issues ranging from equal opportunity, to the environment, to consumer rights. And the judicialization of politics is also mirrored in the dramatic expansion of adjudicating bodies, grievance committees, and other entities that mediate and protect the expansive rights that have cascaded down into private and public corporations.

In contrast to the judicial process, legislative politics is about the collective, public interest, in which commonalities (or at least majorities) are stressed. In emphasizing the "common," the notion of personal space is delimited (to the extreme case of Maoist China, where any form of personal space was viewed with suspicion, including private religion, friendships, and the like). In a collective concept of self (such as in the nation-state) there is less a notion of rights (except to the extent that rights promote engagement in a common body politic), as one is part of a collective whole. To take a graphic example, millions have sacrificed themselves for their nation in wartime. The judicial process is about marking boundaries, not defining communal and national goals. Thus, in the past, courts have deferred to the other branches of government in foreign affairs and other issues, such as immigration, that involve international and cross-border relations. Such judicial intervention was construed as threatening the unitary and sovereign character of the nation, whereas the legislative and executive branches imputedly embodied that unitary, national quality.[44]

The judiciary also serves as a "traffic cop," imposing social order on a world of potentially conflicting spaces; the question becomes all the more critical as "space," no longer territorial per se (in the sense of fixed geographic territories, permanently delimited from one another), has to be constantly negotiated. The judiciary reinforces, or seeks to create, a set of rituals which "respects" social spaces and thus creates or maintains social order. The everyday rules that govern street traffic—neutral and demanding no sense of felt commonality—are, when taken to a higher level, a way of adjudicating a society in which an overarching sense of community is fading. Goffman wrote, using the street example: "Take, for example, techniques that pedestrians employ in order in order to avoid bumping into one another. They seem of little significance. However, they are constantly in use and they cast a pattern on street behavior. Street traffic would be a shambles without them."[45]

This example serves as an analogy for the social order as a whole. What is interesting is that in societies with a strong collective sense of self—a common culture, a rooted collective conscience—personal space may be more contracted, the public squares more lively, the neighborhoods less closed and insular, and suburbanization less marked. Speech is more vibrant, less constrained—everything is akin to a family picnic or squabble. Conversely, where personal space (and group space) is accentuated, the rules of etiquette governing exchanges need to be (for an orderly society) more elaborate, and markers need to be defined. Politeness is the mark of distance. This is all the more significant in multicultural environments. What is of note is that with growing stress on individual and private rights and legally protected cultural expression, the legal framework grows more—not, as one may expect, less—dense, in terms of the number of laws, the kinds of laws, and the caseload of courts and other adjudicating bodies.

Thus, with social space no longer fixed and geographic, other boundary markers come to the fore. This is all the more important in a world where international distinctions have been domesticated and internal (domestic) divisions have become all the more marked. Furthermore, social markers, being fluid and undefined by fixed geographic space, will necessarily involve more "crossings" than geographic borders; in other words, there are many more borders, and those borders are constantly shifting about. As historian of religion Mircea Eliade put it, "The multiplicity, or even the infinity, of centers of the world raises no difficulty for religious [or communal] thought. For it is not a matter of geometrical space, but of an existential and sacred space that has an entirely different structure, that admits of an infinite number of breaks and hence is capable of an infinite number of communications."[46]

"Self-determination" remains important in this context, but, again, it is not necessarily geographically defined. Rather, individual and group criteria come into play in terms of social space ("the territoriality of the self"): "Self-determination turns the whole possibility of using territories of the self in a dual way, with comings-into-touch avoided as a means of maintaining respect and engaged in as a means of establishing regard. And on this duality rests the possibility of according meaning to territorial events and the practicality of doing so. It is no wonder that felt self-determination [in terms of social distance or closeness] is crucial to one's sense of what it means to be a fully-fledged person."[47] By extension, this concept of self-determination can apply to ethnic, social, religious, and other such groups. In relation to "territorial" (in the sense of social space) claims, individuals (and groups) have "the right and duty to call attention to offenses, demanding that something be done to set matters right"—once more highlighting the judicial element.[48]

The growing importance of the language of civil and human rights is part of this: human rights delineate personal and group rights and "spaces" and are central to the international, regional, and national judicialization of politics. In delineating social space, civic and human rights contribute to the ordering of society in the sense of the pedestrian analogy described earlier.

In this playing field, the classic model of civil society and the public-private distinction are turned around: public space is imbued with social distance and avoidance, not community as in the nation-state and the classic civil society model.[49] Community, ironically, is privatized; only in the private realm can social distance be relaxed. Ethnic, social, and religious identities are segmented, apart, personal, and private.

THIS BOOK IS, as previously noted, in two parts, the first dealing with selected American historical times and moments and the second focusing on the contemporary United States. The first part addresses how a people came to nurture a moral association with the land and how the land came to define and represent American nationhood and identity. The second part addresses contemporary issues of what happens when that moral association becomes uncoupled—when the presumption of an inextricable and exclusive tie between a people and the land becomes attenuated. What are the social and political implications of such a development? Threading through the book is an examination of the linkage between bounded territoriality as a marker of nationhood and as central to civic politics and how political forms are fundamentally changed when the social and legal sense of a culturally and nationally homogeneous landscape fades. In considering the marking out of territoriality, nationhood, and politics, law looms large—not only in a narrow legalistic sense but also in forming the sociological weave through which the country defines itself. More broadly, the story told here is about America's changing sense of place.

Chapter 1 considers the Puritans, who, from the early 1600s, first imagined "America." Here we see the beginnings of a moral association of an incipient nation, the Puritans, and a land; a people who conceived of themselves as the New Israel, the Puritans envisioned their colony as the model City on a Hill. The continuity between the Reformation and the Protestant settlement of America is also apparent in discussing this moment in American history. In America, the Puritans believed, the promise of the Reformation, of redeeming this world, would be realized. In part what we see here is how a bounded territory, in the absence of kinship principles, was a sine qua non for establishing "a people" rooted in a civic and republican politics (though one that was theologically closed to non-Calvinists). Contrary to popular perception, the Puritans had a deep sense of being rooted in the land of America, and their view of the "wilderness" was a nuanced one. Scientific thought and development went together with the sacralization of nature. The Puritans' sense of the land was also revealed in repeated discussions about the rights and wrongs of occupying land that may belong to American Indians and when it was legitimate to do so.

Puritan New England was not, of course, the only colony in British America. But the Puritans are, nonetheless, of special interest when considering the themes of place and belonging: The Puritans had a much more defined sense of themselves as creating a new society and a new church and polity; that this was a distinct land divinely ordained for this chosen people sharply differentiated New England from the rest of the colonies on the eastern seaboard in the early 1600s. And, unlike the other colonies, the Puritans sought "not to replicate but to move in precisely the opposite direction of the world they abandoned in old England." The other colonies by and large identified strongly with Britain. Furthermore, and more so than in New England, the other colonies had a strong commercial orientation. It is also equally clear that over time the

South, with its different colonial lineage, developed its own distinctive and competing cultural forms that would ultimately clash with those of the North (and the South is dealt with here, briefly, as a foil to the North, in Chapters 2 and 3).[50]

In the United States after the Revolution of 1776, the starting point of Chapter 2, we see a persistence of the Puritan sacralization of the land. The association of the land with republicanism and nationhood is broadly celebrated in agrarian mythology, which held that civic politics rested in farmers with their own land, independent of external authorities. Like the Puritans, but in a more elaborated form, the land and nature are also celebrated as expressions of God's presence and as sources of economic riches and scientific knowledge. They were to be harnessed in America's progress into the future. So the same seamless weave of land, politics, and nationhood remains. But there are significant shifts, as well. The United States covered a larger swath of territory than New England and was made up by a multiplicity of states. In contrast to the theologically monolithic Puritans, the United States was multidenominational, characterized by a diverse array of sects and churches that represented a more pluralistic and individualistic view of religion. Starting with thirteen states, numerous sects (albeit overwhelmingly Protestant), suspicion of government, and a growing individualism, people generally identified themselves in terms of individual states or smaller geographic units.

We examine in Chapter 2 how the association between people and place was shaped and negotiated, in certain iconic writings and enactments, in the antebellum United States. We consider the writings of Thomas Jefferson, Thomas Paine, and J. Hector St. John Crèvecoeur on the association of the land, people, and republican politics in the United States. Such ideas helped shape institutions, interests, and social possibilities and constraints; we consider the institutionalization of those ideas through definitive events or moments that played a framing role in shaping future developments. In particular, we examine a document of "high, constitutional importance"—mostly overlooked today—the Northwest Ordinance of 1787. That ordinance played a central role in choreographing American expansion westward. In this light, the sectional struggle over slavery and the distinct notions of place, space, and time which characterized North and South respectively are considered.

In Chapter 3 the conservation and preservation movement, which arises in the post–Civil War period, is discussed. It is in this period that the national parks and monuments were established—significantly, because it is in this period that the United States is presented as an unambiguously unitary land and nation. This chapter analyzes this movement in the context of the nationalization of American identity after the collapse of the South in the Civil War. War monuments and memorials, in particular, starting with Abraham Lincoln's Gettysburg Address, forged the link between civic sacrifice and nationhood. In the very act of national mourning, in shared melancholy, civic sacrifice and unity, nation and soil, the singularity of the people and the land was engendered. The public square, now the figurative and in part literal embodiment of civic life in a more urban America, was anchored in war monuments, often set in a "natural" scene of trees and fauna—representing civic sacrifice, the mingling of the blood of lost soldiers with the soil, and nature.

The sense of the land as the organic body of the nation had been expressed in the landscape painting of the Hudson River School through much of the nineteenth century. Lincoln took that already existing northern sentiment and sketched the landscape as the bedrock of an egalitarian and united Republic. He compared slavery to the ascriptive, kinship principles of monarchies. In contrast to Jefferson's farmer in immediate and palpable touch with the land, national parks and monuments become important in mediating the link with the land and nature for a now national and increasingly urban community. Prior to the Civil War humanity and nature had a symbiotic relationship (what has been called the Middle Landscape)—men and women *belonged* in nature. Now, in the face of industrialization and urbanization, nature was depicted as pristine only when free of human intrusion and development, and it was in that milieu national parks were established. Now citizens made pilgrimages, through mass tourism that began in the late nineteenth century, to national icons and parks.

The importance of the parks and monuments in defining the contours of American civic and political life led to their becoming, over time, battlegrounds for defining the very nature of American society. In Chapter 4 we ponder how, in the past three to four decades, public monuments and memorials were replaced, reinterpreted, or supplemented by other icons and markers. Indeed, the relandscaping of monuments, museums, parks, and other sites suggests far-reaching change in the meaning of "place" in America. This is especially evident in the changing representation of violence (war monuments and memorials to events related to violent struggle being the most prominent form of public commemoration). Traditional monumental symbols of nationhood may "harden," generating certain nonchalance. Rather than monuments of time eternal, they represent the deadening weight of an antediluvian past and as such may even generate hostility. Those symbols thus become the targets of rebellious and reformist change. Such an environment, questioning the past and its public representation, became broadly visible starting in the 1960s and 1970s. A case in point is the American Indian Movement, which initiated protests in that period about the Custer Monument in Montana. Such protests asked what kind of America was being celebrated, what its social contours were, and whose moral ties to the land should be noted.

As a result of these struggles over the organization of public space and public commemoration, patterns of memorializing changed in the United States. A series of parks and monuments were established, or older parks were redefined, as "ethnic heritage" or to express national atonement for past wrongs, or were described as women's history or women's rights parks. The National Park Service offered new "interpretations" of history which moved away from the heroic description of American history to a more "multiple perspectives" or multicultural approach. The most prominent national parks, such as the Grand Canyon, were declared World Heritage Sites, under the auspices of the United Nations. The sensibility of a bounded nation with an exclusive moral tie to the land, or a culturally homogeneous landscape, was no longer presumed. Violence itself came to be interpreted in vastly different ways, particularly at two highly visited sites in Washington, D.C., the Vietnam Veterans Memorial and the Holocaust Memorial Museum. In Chapter 4 we consider the Little Bighorn National Monument (formerly the Custer Monument), the Vietnam Memorial, the Holocaust Memorial, and the Grand Canyon National Park in examining America's changing moral geography.

Spatial patterns have a temporal rhythm—they organize the sense of time and movement. The "public square" of traditional civic politics metaphorically suggested the nation writ small—bounded, enclosed, rational, and linear in time and space. In Chapter 5, we consider how the "spatialization" of the contemporary United States is changing in contrast to the civic and republican vision. In a country that, since the 1990s, is mostly suburban, spatial and social patterns are much more fluid, mobile, and privatized. Cultural markers are malleable, voluntary, and not fixed geographically. Courts and administrative entities play a more salient role in mediating these social spaces, spaces that are defined though the idiom of rights. Law—through court cases, administrative rules, and the like—becomes denser as the "territories of the self" get ever more legally protected; this in turn expands the legal "agency" of the individual to determine his or her cultural and ethnic self-expression. In contrast to the "public square" we have a spatial rhythm more akin to pilgrimage, whereby the sacred, or symbolic, centers may be located elsewhere and movement is favored over stasis. This is reflected in the rise of transnational communities and in the patterns of migration and tourism, as well as in modes of association facilitated by new information technologies. And it is a pattern in which public spaces are largely absent of civic or political activities and communal affiliations are a private matter—they become privatized public spaces, such as shopping malls.

Chapter 5 also extends this analysis of the quilted texture of the landscape, of the changing configuration of social and territorial spaces, by considering in particular how the inclusion of immigrants in American society has changed. In the past the incorporation of immigrants was predicated on their "fit" into a landscape said to reflect a singular and homogeneous culture—and this was the case for liberals as well as restrictionists on immigration matters. Now, as themes of cultural homogeneity and exclusiveness lessen in their imprint, the courts seek to mediate a more multicultural environment, where felt identities may be diasporic or multicultural and where assimilation is not presumed. At the same time the ground rules that are instituted to support multiculturalism—cultural identity cannot be imposed on a person—constrain it. When a culture has elements that are not based on individualistic and voluntary principles—say, requiring arranged marriages—such elements are "edited" out, or the culture is reinvented, to fit a multicultural society. In this process, there is a curious interplay of cultural diversity with a homogenization of culture, as reflected in privatized public spaces, where different cultural expression meets. The chapter concludes with a brief contemplation of an intriguing vista: America as the home of a polyglot of cultures in a world increasingly home (if at times reluctantly) to things American, cultural and material.

In the Coda we reflect, in a more philosophical vein, on the nature of place in human life—its reconciling, or seeking to reconcile, the individual, the nation, the people, in the finite "here and now" with the infinite, incomprehensible torrent of space and time.

An American Eden

ONE

But now we are all, in all places, strangers and pilgrims, travellers and sojourners: most properly, having no dwelling but in this earthen tabernacle. Our dwelling is but a wandering; and our abiding, but as a fleeting; and, in a word, our home is nowhere but in the heavens; in that house not made with hands . . .

Now such should lift their eyes and see, Whether there be some other place and country to which they may go, to do good . . . which GOD *hath given them for the service of others, and his own glory?*

—ROBERT CUSHMAN, 1622

AMERICA, like everything else in human life, had to be imagined before it could be seen (even by those who were already there). And it was the Puritans who first imagined America, not as a colony, not as an expansion of a European or British enterprise, but as the wilderness, the landscape in which humankind would be reborn. In America humanity would, as God's steward, begin anew. In America, the Reformation would be completed in an apocalyptic climax. In America, politics would become the handmaiden of a godly mission, in which civic activities were a collective calling. The Land had been "spied" out by God, a Promised Land for a chosen people with a revolutionary political and civic vision. The synthesis of land, people, and civic politics would contribute greatly to the emerging idea of the nation-state and channel the course of history, not only of America, but of the world.

The roots of America as an idea lie, historically and in the way the early

Americans imagined themselves, in the Protestant Reformation itself, by way of its initial success, then failure, in England, and in the "discovery" of the New World as the soil, the Promise, the place where Redemption would be found far from the now corrupt England. The Reformation offered a radical reorientation to perhaps *the* central markers of human life—movement, space, and time. *Place* is the intersection of time and space, and place is demarcated through patterns of movement and settlement. The Germans, the Dutch, the English, and others would now view themselves as sacred agents in history, their very soil blessed by God as they strove to bring on the End of History. As Avihu Zakai writes, the "Protestant Reformation was situated at the end of time and history, as an eschatological event preceding that moment when the whole mystery of providential history was to be resolved."[1] That palpably near sacred moment, when the here and now became the All in All, took place *here,* in England, in Scotland, in Holland.

The medieval world was, in contrast, passive, local, ahistorical, and the "world" was a patchwork of quilted loyalties and identities. Life was truly mundane, in the religious sense, broken perhaps by pilgrimages to sacred centers and shrines such as Santiago de Compostela, Rome, and, for a tiny few, the Holy Land itself. The "here" was not very stirring or promising; it was the endless mill-like routine of life, where one day rolled meaninglessly into the next, where salvation would have to wait for, literally, another world. Under Protestantism, England, Holland, and other Protestant countries became the Holy Land, the New Israel, driving forward history and time. They each viewed themselves (often in tandem with other Protestant peoples) as having this providential role. The land and history, as well as the people, became intertwined. England was this "precious stone set in the silver sea," and one could imagine that the "Countenance Divine" did "Shine forth upon [England's] clouded hills."[2]

But only for a period of time. For the Protestant "purists," the English failed the Reformation. Protestantism in its first two hundred years in England, it must be remembered, was often on the defensive and in control only through bloody suppression. Mary, who became queen in 1553, beat back Protestant gains of the mid-sixteenth century made under the reigns of Henry VIII to his son Edward VI. Mary, daughter of Henry VIII and half sister of Edward VI, burned notable Protestants at the stake in her bid to restore Catholicism in England. She consequently gained the sobriquet "Bloody Mary." When Elizabeth ascended to the throne after the death of Mary in 1558, she sought a middle road politically—a path of compromise that was to become the hallmark of English politics—between Catholicism and the "purist" Protestants (who would receive the name "Puritans" only in the 1560s). The Elizabethan settlement was a difficult—indeed, impossible—compromise for the Puritans, as it represented for them a reversal of history, a step back from the rush onward to the apocalypse, and a betrayal of England's sacred and providential role. The very symbols and ritual that the Church of England had inherited from Catholicism were an affront to the purists' desire to rid the church of "papist" corruption, dogma, and ritual. The elaborate stained-glass windows, statues, and crucifixes symbolized the mediative and hierarchical role of the Catholic Church, in contrast to the austere Calvinist churches, where nothing came before the saints and their God. In the wake of the Elizabethan compromise, Puritans began to organize, agitate, and seek to reignite the Reformation. In this context, England was desacralized, stripped of its providential role; more than that, England in its degeneration had strayed into the path of Satan.[3]

English Anglican (Church of England) settlements, such as Virginia, when imbued with religious purpose, were viewed as the expansion of an elect England in its glorious, providential role in salvation. Many Puritans, on the other hand, were now (by the early seventeenth century) so deeply alienated from the Church of England that they viewed it as the Antichrist; they saw America as God's Promised Land and refuge for his newly chosen people, the place where the Reformation would be realized. England, if it continued to, in the words of John Winthrop, "imytayte Sodom in her pride and intemperance" and to be "the Sinogogue of Antichrist in her superstition," would be scorched by the wrath of God.[4] Winthrop, a leading figure among the Puritans who would play a central role in shaping church and state in New England, stated further, "I am veryly perswaded, God will bringe some heauye Affliction vpon this lande [England]." But, Winthrop wrote to his wife, "the Lord . . . will prouide a shelter and a hidinge place for vs."[5] At this time the Puritans felt doubly beleaguered, as Protestant churches all over Europe were in retreat; the Huguenots had been vanquished in France, the Catholic Church had been restored in regained ecclesiastical estates that had been lost since the Peace of Passau in 1552, and the Lutheran Danish king Christian IV had withdrawn from the Thirty Years' War.

America became more attractive as, in the words of one Puritan, "the more that wee see the moste parte of the protestante churches of Europe are destroyed."[6]

Here we witness the conception, if not the birth, of the "first new nation." It is a remarkable development on a number of levels. We see the sacred (if, at this point, projected and imagined) joining of a people and a land. And, as discussed below, it is a "joining" that fundamentally and intrinsically involved a "civic" polity—and so we begin to see the interweaving to the point of a "self-evident truth" of community and civic polity with a fixed and bounded territoriality. For the Puritans, this idea was captured in the notion of Covenant, which bound together God, land, and people, and the people with one another. We see in this experiment also the genesis of a nascent nation-state; if the Reformation broke western Europe up into multiple, sovereign territorial states, Puritan New England was going to be a model, a "City upon the Hill," for an emerging international order. Although not the first expression of an emerging conception of a territorially constituted people with a historical, even apocalyptic history—the Dutch republic may claim that status in terms of modern history—the Puritans anticipated and would help give transcendent and sacred meaning to the settlement that brought the modern system of sovereign states into existence, the Treaty of Westphalia of 1648.

The continuity between the Reformation and the Protestant settlement of America was a theme that remained in America well after the first-generation Puritans passed on (or for that matter, after Puritanism itself had disappeared in the American landscape). "The Reformation was preceded by the discovery of America," Tom Paine wrote in 1775, "as if the Almighty graciously meant to open a sanctuary to the persecuted in future years."[7] The settlement of America and, later, the Revolution were viewed as a sequence of events in God's design. "I always consider the settlement of America with reverence and wonder," wrote John Adams in 1765, "[and] as the opening of a grand scene and design in Providence for the illumination of the ignorant, and the emancipation of the slavish part of mankind all over the earth." Jonathan Edwards pronounced that "our nation" is "the principal nation of the Reformation." America, the New Israel, was chosen to do God's work, to push forward the course of history. This struggle would ultimately be directed for causes such as abolitionism and spreading democracy.[8]

But, before we get ahead of ourselves, let us first go back to that time when a people felt the frisson, the exhilaration, of reaching the cardinal moment in history, when God had appointed a people—themselves, naturally—to bring on the millennium and had chosen the land sacralized for that undertaking, America. It is at this point that the moral association between people and place, between America and a nascent American people, is first articulated. And it is a moral association that anticipates, and serves in part as a precedent for, a nationalist *mentalité* the world over. This special association is articulated in a series of sermons in the years that preceded and followed the "Great Migration" to New England in 1630.

What is striking about these sermons is the sense of being "dis*placed*," with no point of orientation to the world. The world is in the sense of Genesis a moral void. Travel is a travail, and we are all refugees. Robert Cushman, in a sermon dating from 1622 (one of the earliest in this regard), *Reasons and Considerations Touching the Lawfulness of Removing Out of England into the Parts of America,* pictures a landscape where the people are wandering, lost. With no point of orientation, no distinction between sacred and profane, there is no right and wrong, no direction. It is only in removing oneself to America that one will have "place" and be located once more, historically, geographically, and, above all, morally. It is akin to the People of Israel, lost in Sinai, finding the Promised Land:

> Neither is there any land or possession now, like unto the possession which the Jews had in Canaan; being legally holy, and appropriated unto a holy people, the Seed of ABRAHAM: in which they dwelt securely, and had their days prolonged. It being by an immediate Voice said, That he, the Lord, gave it to them, as a land of rest after their weary travels; and a type of eternal rest in heaven.
>
> But now there is no land of that [sanctity]; no land, so appropriated; none, typical: much less any that can be said to be given of GOD to any nation, as was Canaan; which they and their seed must dwell in . . . But now we are all, in all places, strangers and pilgrims, travellers and sojourners: most properly, having no dwelling but in this earthen tabernacle. Our dwelling is but a wandering; and our abiding, but as a fleeting; and, in a word, our home is nowhere but in the heavens; in that house not made with hands.

Keeping in mind a sense of *place* where time (history) and space (location) interact, and where the here and now is enveloped by the All in All, where "we" are filled with God's purpose, where, to use nineteenth-century language, the Kingdom of Heaven on Earth was at hand, that place was America: "Now such should lift their eyes and see, Whether there be some other place and country to which they may go, to do good . . . which GOD hath given them for the service of others, and his own glory?"[9] It would be decidedly un-Puritanical, given the Puritans' preference for sober and methodical disquisition, but one can almost hear the enthusiasm of an evangelical sermon and the rhythmic cadence of Martin Luther King Jr. singing out, "And he has allowed me to go up the mountain. And I've looked over. And I've seen the promised land. I may not get there with you, but I want you to know tonight that we, as a people, will get to the promised land. And I'm happy tonight. I'm not worried about anything. I'm not fearing any man. Mine eyes have seen the glory of the coming of the Lord."

America becomes a home from the storm, a home built by God. There is that curious tension in Protestantism, particularly in worldly, politically minded Protestants like the Puritans, between the desire for *movement,* to progress historically in the drive to the millennium, and the desire to *order* the world, to conquer the earthly kingdom. This dynamic was also married to the larger Christian story of salvation, from the alienation from Eden to reconciliation with God, from the escape from Egypt to the pilgrimage to the Promised Land.[10] We are journeying to the Promised Land, metaphorically, in the sense of some Edenic future. But that Edenic future is also located physically; we are journeying to America, where the future will be realized. The journey would not be an easy one; it was an escape from corruption and sin, a testing of one's soul like the wanderings of the Jews in Sinai, for the promise of America. But in America the journey would begin again, to create a godly society.

And so the myth of immigration became part of the very sinews of America's conception of its collective self, of its soul.[11] To be rooted was contrasted with the "transient," wearying quality of travel. How can we be but captured by the awe of the poet Thomas Tillam, as he viewed the new Canaan (in his poem *Uppon the first sight of New-England, June 29, 1638),* contrasted with the arduous voyage to reach it:

Hayle holy-land wherein our holy lord
Hath planted his most true and holy word
Hayle happye people who have dispossest
Your selves of friends, and meanes, to find some rest
For your poore wearied soules, opprest of late
For Jesus-sake, with Envye, spright and hate
To yow that blessed promise truly's given
Of sure reward, which you'd received in heaven
Methinks I heare the Lambe of God thus speake
Come my deare little flocke, who for my sake
Have lefte your Country, dearest friends, and goods
And Hazarded your lives o'th raginge floods
Posses this Country; free from all anoye
Heare I'le bee with you, heare you shall Injoye
My sabbaths, sacraments, my ministrye
And ordinances in their puritye
But yet beware of Sathans wylye baites
Hee lurkes amongs yow, Cunningly hee waites
To Catch yow from mee; live not then secure
But fight 'gainst sinne, and let your lives be pure
Prepare to heare your sentence thus expressed
Come yee my servants of my father Blessed.[12]

William Bradford's *History of Plymouth Plantation* blends historical narrative and theological exegesis in his description of the harrowing voyage of the Pilgrims to Plymouth Plantation. In his history of the Pilgrims the Exodus theme comes through explicitly, of the "wandering in the wilderness" before the Israelites—now the Pilgrims—find their *place,* their home. Calvinist history was designed to reveal divine intentions, and it had a didactic purpose. New England was the apogee of the Reformation, at the end of an arduous voyage. William Bradford was a Separatist, distinct from, if theologically similar to, the Puritans, part of a group that, unlike the Puritans, completely split off from the Church of England. This group, more familiarly known as the Pilgrims, escaped to Holland in 1609 (the year the Dutch won their independence during the throes of the Dutch Revolt, the Calvinist rebellion against the forces of the Catholic Counter Reformation). The first Pilgrims left for America, ending up in Plymouth, in 1620, and Bradford was elected governor of the colony in 1621.[13] In a passage almost beseeching the Lord while expressing gratitude, Bradford wrote:

> Our faithers were Englishmen which came over this great ocean, were ready to perish in this willdernes; but they cried unto the Lord, and he heard their voyce, and looked on their adversitie, etc. Let them therfore praise the Lord, because he is good, and his mercies endure for ever. Yea, let them which have been redeemed of the Lord, shew how he hath delivered them from the hand of the oppressour. When they wondered in the deserte willdernes out of the way, and found no citie to dwell in, both hungrie, and thirstie, their sowle was overwhelmed in them. Let them confess before the Lord his loving kindnes, and his wonderfull works before the sons of men.[14]

And so from the sense of loss, of place and of direction, in a world gone amok and dissolute, expressed in Cushman's sermon and in Bradford's history, to the decision to take the arduous, difficult journey to Redemption, we come to America. It was the first day of Genesis. And there was light across New England, like some nineteenth-century painting of the American landscape. The world could now be ordered—in contrast to the void and chaos that preceded the first day of Genesis. The landscape revealed Providence, his sacred presence, and a moral order to the world. In America, the saints, as Tillam wrote, could realize divine ordinances and fulfill their covenant with God. The people had discovered their place, literally, morally, metaphysically, and historically.

The passage from 2 Sam. 7:10—"Moreover, I will Appoint a Place for my people Israel, and I will PLANT them that they may dwell in a place of their OWN, and MOVE NO MORE"—served as the basis of a sermon (the source of the capitalization noted in the quotation) given by John Cotton in March 1630 as the Puritan flock gathered in the port of Southampton, ready to set sail to New England. Cotton was a clergyman in England before fleeing to New England in 1633 after being summoned to court for his Puritanism.[15] His sermon at Southampton, *God's Promise to His Plantations,* is perhaps the most notable and famous of the tracts that justified the settlement in America, on the grounds of a special moral association between that land and this people. Staking out the land of New England as a distinct sovereignty separate from England, Cotton was also clearly making a set of moral claims. What is also notable about this tract is its implicit picturing of the world as distinct territorialities tied to distinct peoples, all at God's behest. New England and Yahweh's new chosen people had, of course, an especially significant place in this constellation. Cotton sermonized:

> This placing of people in this or that Country, is from God's Soveraignty over all the Earth, and all the Inhabitants thereof . . . in *Jer.* 10.7 God is there called *The King of Nations* . . . Therefore it is [meant that] he should provide a place for all Nations to inhabit, and have all the Earth *replenished.* Onely in the Text here is meant some more special Appointment, because God tells them it by his own mouth . . . That is, He gives them the Land by Promise. Others take the Land by his *Providence,* but God's people take the Land by *Promise:* and therefore the Land of *Canaan* is called a Land of Promise.

This is a remarkably revealing statement about Puritan conceptions of the world (at least as the world ought to be) but more so of an incipient conception of what would come to be called "nation-states" and even "international society." This statement precedes the Treaty of Westphalia of 1648, which established the institutional framework of sovereign states, by a generation and illustrates the importance of the Protestant Reformation in what later would be called the (nation-) states system. Furthermore, Cotton stresses the importance of *bounded* territorialities. Drawing on the verse from Samuel, he notes, "The placing of a people in this or that Countrey, is from the Appointment of the Lord," and states further: "*God hath determined the times before appointed, and the bounds of our habitation* . . . God would not have the *Israelites* meddle with the *Edomites,* or the *Moabites,* because he had given them their Land for a possession. God assigned such a Land for such a Posterity, and for such a time." Here one senses a certain realpolitik, that territorial demarcations are a device for engendering peace. This is not surprising, given the intensely political character of the Puritans and their readiness to engage, if necessary, in combative struggle—a realism and quality that would be evident in later American figures such as Alexander Hamilton. (We shall return to the "boundedness" of territorialities later.)[16]

Cotton's immediate concern, however, was to evoke the sacredness of New England's soil for the chosen people, the Puritans. Cotton draws a direct relationship between God's promise to the Israelites of the Promised Land and the journey the Puritans were about to undertake to New England. God himself had chosen this place and had chosen "to plant" his people, the Puritans, there. The metaphor of planting is significant: the people would come of the soil, "as a Plant sucks nourishment from the soyle," and it would be their place. God will plant his newly chosen people in their "*Own* Land" and cause them to grow as plants do. In an allusion to the Covenant, when the "planted" people become "Trees of Righteousness," so the Lord himself is planted. In a sentence that is remarkable in its display of the special moral association between America and the New Israel, the Puritans, Cotton writes of God planting "us" in his "holy Sanctuary [but in doing so] he is not rooting us up." The people were, so to speak, native plants; the soil was their soil, the land their land. Cotton has a warning, as well: "If God be the Gardiner, who shall pluck up what He sets down?" But, he notes, a number of paragraphs after this rhetorical question, "While they continued *God's* plantation, they were a noble Vine, a right Seed: but if *Israel* destroy themselves, the fault is in themselves."

Cotton was also alluding to Gen. 2:18: "And the Lord God planted a Garden eastward in Eden; and there he put the man whom he had formed." God *places* man and conversely punishes him by sending him into exile as a wanderer. Man himself is made of the soil—"the Lord God formed man of the dust of the ground, and breathed into his nostrils the breath of life; and man became a living soul" (2:7). The New Israel would dwell there, Cotton said of God's blessing, like freeholders; their possession of the land would be "firm and durable," with no need to move again; and they could rest peacefully there. The "sons of wickedness" could no longer afflict them in the New World. Nevertheless, "Neglect not Walls, and Bulwarks, and Fortifications for your own defence; but ever let the Name of the Lord be your strong Tower; and the word of His Promise the Rock of your Refuge. His Word that made Heaven and Earth will not fail, till heaven and Earth be no more."

This land, furthermore, was unlike any other—other lands were allocated, as noted earlier, to other peoples by Providence, but this was the Promised Land. American soil was sacred, tied by Covenant to the Puritans as they were to God, an axial point in the rush to historical redemption. God, said Cotton, had "spied" out the land and lets his people discover it or hear of its discovery. This is their own country, not a colony as such. Having spied out the country God, "carries [his people] along to it, so that they plainly see a providence of God leading them from one Countrey to another: As in *Exod.* 19.4. *You have seen How I have born you as on Eagles wings, and brought you unto my self.*" Cotton continues, "So that though they met with many difficulties, yet he carried them high above them all, like an Eagle, flying over Seas and Rocks, and all hinderances."

Cotton is clearly deeply concerned about the legitimacy of this claim of Puritan proprietorship over the land, and he does not end his argument purely on, so to speak, God's word. As we see later in this chapter, he spends some time explaining the question of possibly displacing those who are native to America or, specifically, to New England.

Other prominent Puritan saints made similar, if not so eloquent, arguments, stressing, among other things, that God was providing a place as a refuge in the New World for his people before he unleashed his wrath on the depraved and corrupt English and other Europeans.[17] Thus began the theme, still present, of America as an asylum and refuge.

Civic Politics

Calvinists, in many cases the soldiers of the Protestant Reformation, played a critically important role in the emergence of civic politics.[18] The Puritans in America were no exception—indeed, they were exemplars of a certain kind of civic community. If anything about Puritan political documents remains in the popular, certainly American, memory, it is the Mayflower Compact of 1620 and John Winthrop's sermon on board the *Arrabella* on the way to New England in 1630, "A Modell of Christian Charity." Both these documents describe a desire to create a civic community based on Christian theology and self-government. Less noticed, however, is the fundamentally territorial nature of these civic polities. This is significant because it is here we begin to see how a certain kind of politics was associated with a certain kind of peoplehood (or nationhood) and how both such an identity and politics were rooted in the Land. It also shows how we came to the point where, when we said "America," or "England," and so on, we could unconsciously treat community, polity, and physical place as one, unproblematic unitary and seamless entity.

The "Great Migration" that took place in the spring of 1630 consisted of eleven cargo boats, seven of which had been converted to take passengers on the three-thousand-mile voyage across the Atlantic. These boats carried seven hundred people, drawn from all classes, from a handful of aristocrats to working people, mostly tradespeople, artisans, and yeomen but also servants and laborers. They came from all over England. As the boats crossed the Atlantic, John Winthrop delivered, most likely by voice, his sermon "A Modell of Christian Charity." In turning their back to England, they had not only cast away their affiliation and loyalty but had also forsaken England's social institutions. They did not carry with them the assumptions of government determined by divine right and kinship, nor did they accept hereditary rank. They had to begin anew, politically and socially. In his sermon Winthrop articulated a new, civic form of association, uniting the people voluntarily on the basis of shared theology and Christian purpose. Government, ecclesiastical and civil (deeply intertwined, such that the state served the church), was to embody and represent the faith and wisdom of the Puritan saints. In this sense, they had a covenant among themselves—a civic obligation—to create a society formed by Christian beliefs. They also had a Covenant with God; God had spied out the Land for them, made space for them, and had chosen them to be his people. They, in turn, had to live according to, and to realize, God's "ordinances," his law.

It is in the very creation of a civic sphere—a "public"—that Puritan social and political identity was to be realized. In the absence of kinship or blood descent, which hitherto had determined social rank, rulership, and place in society, legal criteria—explicitly defined moral codes—came to play a critical role in defining the community. That public sphere was defined in legal terms. In the case of the Puritans, that law was God's law, and the language of living and realizing God's "ordinances" abounds in Puritan writings. More striking in this context was the idea of the Covenant, the cornerstone of Puritan society. The Covenant was, in essence, a legal contract among the Puritans themselves and between the saints and their God.[19] The term *covenant* is derived from the Hebrew *berit* (also signifying the Israelite covenant with Yahweh), which translates as alliance or agreement and is, at least implicitly, a legal understanding. The Puritan Covenant essentially *constituted* the civic sphere, playing the same role that the Constitution later would, albeit the latter in a more secular form.

The civic or public sphere, moreover, was dynamic in that it captured a sense of history, of movement, of a drive towards some apocalyptic or millennial future (or utopia, in the case of some nationalist movements). The Puritans were an intensely worldly, political community; unlike some pacifist and mystical Protestant groups that had preceded or were contemporaneous with them—groups that sought to unite mystically with God and viewed this world as incorrigibly corrupt—the Puritans saw political, worldly control and conquest as critical in God's divine plan, in bringing on the End of History and the heavenly kingdom on earth. In that light, the civic girding of the saints was a critical event, part of the political process of realizing the Reformation. In the same way, civic politics was also an inherently territorial phenomenon, as it involved the need to order the world, to delineate and determine God's jurisdiction territorially, and the territorial jurisdiction of his people. Civic politics was the form of, and formed, communal self-definition.[20] People and civil polity become synonymous with "place" and constituted, and are constituted by, a "territory." In the absence of kinship, and with presumptions of equality (within the band of saints), communal and collective identity comes to be defined civically and territorially. Collective identity, civic polity, and territory are completely intertwined.

"A Modell of Christian Charity" and, to a lesser extent, the Mayflower Compact, as well as other sermons and declarations of the time, are illustrative in this regard. In Winthrop's address, he described the traits of a civic society:

> It is by a mutuall consent, through a speciall overruleing providence and a more than an ordinary approbation of the Churches of Christ, to seeke out a place of cohabitation and Consorteshipp under a due forme of Government both civill and ecclesiasticall. In such cases as this, the care of the publique must oversway all private respects, by which, not onley conscience, but meare civill pollicy, doth binde us; for it is a true rule that perticular Estates cannot subsiste in the ruine of the publique.
>
> The end is to improve our lives to doe more service to the Lord; the comforte and encrease of the body of christe, whereof wee are members; that ourselves and posterity may be the better preserved from the Common corrupcions of this evill world, to serve the Lord and worke out our Salvation under the power and purity of his holy Ordinances.[21]

In New England, the Puritans would have to create their community out of whole cloth; it would be voluntary and, Winthrop suggests, created out of shear religious will, on the basis of Christian principles. (The Mayflower Compact touches on a similar theme, but it is much briefer than Winthrop's address.) It is through the creation of a civic society—a common public—that "community" would be effected and that civic society 's purpose, reason, and definition would be the realization of God's ordinances. Furthermore, Winthrop is explicit in noting that the public is more than the sum of the parts and more than a consensus derived from complementary interests. The civic sphere becomes the very source of identity and reward, akin to love—be it marital love or, more important, love of God, which he compares to the bond of marriage. "If any shall obiect," stated Winthrop, "that it is not possible that loue should be . . . upheld without requitall . . . [they should note] this love is allwayes under reward it never gives." What we have, then, is the organizing of a public "persona" and a stress on the collective over the individual interest, so as to serve the Lord:

> That which the most in theire Churches mainetaine as truthe in profession onely, wee must bring into familiar and constant practise; as in this duty of love, wee must love brotherly without dissimulation, wee must love one another with a pure hearte fervently. Wee must beare one anothers burthens. We must not looke onely on our owne things, but allsoe on the things of our brethren . . .
>
> . . . Wee must delight in eache other; make other's Condicions our owne; rejoice together, mourne together, labour, and suffer together, allwayes haueing before our eyes our Commission and Community in [our] worke, . . . soe shall wee keepe the unitie of the spirit in the bond of peace. The Lord will be our God, and delight to dwell among us, as his owne people, and will commaund a blessing upon us in all our wayes.

When Winthrop turns to the Covenant, it is clear that the civic covenant was not only between the Puritans themselves but with God and that the Covenant called for an active role by the saints, socially and politically (or civically, which captures the interlinked nature of the "social" and the "political"). In New England, the Puritan historian Perry Miller wrote, "the specific terms of a compact between God and man . . . rested . . . not upon mere injunction but upon a mutual covenant in which man plays the positive role of a cooperator with the Lord."[22] Winthrop warns:

> When God gives a speciall commission he lookes to have it strictly observed in every article; When he gave Saule a commission to destroy Amaleck, Hee indented with him upon certain articles, and because hee failed in one of the least, and that upon a faire pretense, it lost him the kingdom, which should have beene his reward, if hee had observed his commission. *Thus stands the cause betweene God and us. We are entered into Covenant with Him for this worke* . . . Wee have hereupon besought Him of favour and blessing: *Now if the Lord shall please to heare us, and bring us in peace to the place we desire, then hath hee ratified this Covenant and sealed our Commission,* [and] will expect a strickt performance of the Articles contained in it; but if wee shall neglect the observacion of these Articles which are the ends wee have propounded, and, dissembling with our God, shall fall to embrace this present world and prosecute our carnall intencions, seeking greate things for ourselves and our posterity, the Lord will surely breake out in wrathe against us; be revenged of such a [sinful] people and make us knowe the price of the breache of such a Covenant.[23]

Thus the very opening up of New England was part of the covenantal promise to God; the Puritans had an "errand into the wilderness" to "vindicate the most vigorous ideal of the Reformation, so that ultimately all Europe would imitate New England."[24] The territorial and legal character of civic politics is consequently broached by Winthrop in his sermon on the *Arrabella*:

> [The Lord] shall make us a prayse and glory that men shall say of succeeding plantacions, "the Lord make it likely that of New England": for wee must Consider that wee shall be as a Citty upon a Hill, the eies of all people are uppon us; soe that if wee shall deale falsely with our god in this worke wee haue undertaken, and soe cause him to withdrawe his present help from us, wee shall be made a story and a by-word through the world . . . *Wee are commaunded this day to love the Lord our God, and to love one another, to walke in his wayes and to keepe his Commaundements and his Ordinance, and his lawes, and the Articles of our Covenant with him that*

> *wee may live and be multiplied, and that the Lord our God may blesse us in the land whither wee goe to possesse it:* But if our heartes shall turne away, soe that wee will not obey, but shall be seduced, and worshipp . . . other Gods, our pleasure, and proffitts, and serve them; it is propounded unto us this day, wee shall surely perishe out of the good land whither wee passe over this vast sea to possesse it.

Not surprisingly, given the high hopes, within a generation after the settlement in Massachusetts Bay, the Puritan divines were bewailing civic decline—the start of a long tradition in American history, lately taken up by sociologists and political scientists. By the 1660s jeremiads appear castigating the people for failing to keep the Covenant. Interestingly the afflictions God is said to rain down on the people are, in good part, natural disasters such as epidemics, crop failures, invasions of grasshoppers and caterpillars, hellish summers, and arctic winters.[25] It is as if the people, in their civic failures, had defiled the land and were now almost literally reaping the consequences. The sins representing such decline were the kind that England had previously been fingered for: murder, slaughter, incest, adultery, "whoredome," and drunkenness. Sloth and idleness, it was suggested, were the nursery of all evil in the commonwealth and would hasten the ruin and dissolution of the body politic.[26] There were other signs of decay, as well, some still with us: people were becoming more contentious, such that lawsuits were growing and lawyers were happily prospering from the discontent. Licentiousness was rife—"women threw temptation in the way of befuddled men, by wearing false locks and displaying naked arms and legs or, which is more abominable, naked Breasts," and the "bastardy" rate was going up.[27]

The linkage between a civic polity, law, and creating communal identity through the realization of a "public" was thus innately of a historical and a geographic form. Consequently *place* came to represent in a singular fashion different levels—communal, political, geographic, and historical—of being.

The centrality of law in the constitution of society had other implications, as well. As the sociologist Steven Grosby writes, law becomes the vehicle for the dispersion of the power of the deity. God's presence is defined legally, which not only joins people and land but also ensures its *bounded* character. Boundaries mark the jurisdiction of the law of God, of his people and their land. In medieval society, the sacred was contained in specific sites. These sites represented the *axis-mundi*, where God's finger, so to speak, touched the earth, such as the sites of venerated saints, holy temples, the Virgin Mary, holy waters that cured ills, and so on. One did not inhabit the sacred, as the Puritans did in New England, and as other such religious and nationalist movements would view themselves as doing, but one journeyed to those sacred points. Medieval "borders" consequently were much more amorphous, with more provincial and mundane connotations, and thus borders tended to overlap and have various levels of significance and were more easily crossed than contemporary borders. In the medieval world the sacred was, in a manner of speaking, "vertical," represented by a hierarchical church designating points of mediation with a holy spirit. For the Puritans, and the societal form they propagated and which was to become global in its import, the sacred was "horizontal" and bounded. Inhabiting the sacred, one did not need to be at a particular locality to experience grace.[28] It is no surprise, in this context, that the Protestant reformers condemned pilgrimage.

Cotton elaborated on the role of law in delineating the New Israel, using the metaphor of "planting" simultaneously with regard to the new chosen people and the law. "Have care," he admonished, "to be *implanted into the Ordinances*, that the word may be ingrafted into you, and you into it," and "God plant[ed] his Ordinances among you."[29] The people's sovereignty lay in upholding the law—"you shall never find that God ever rooted out a People that had the Ordinances planted amongst them, and themselves planted into the Ordinances." Elsewhere Cotton notes, again linking the Land to law, "When God wraps us in with his Ordinances, and warms us with life and power of them, as with wings, there is a Land of Promise."[30]

Grosby notes how in the case of ancient Israel—a political and sociological precursor of the Puritans as well as a theological model—religious laws such as kosher practices applied throughout Canaan in order to uphold the sacred purity of the land. As such, as Grosby writes, territory is a structure of "bounded, spatial communion." The land is personalized, becoming the locus of historical memories and future hopes and expectations. It is conceived of as in "moral sympathy" with the people—it is their *country*, much as a house becomes, in a moral sense, a home.[31] Such a sense of territoriality became the sacred and conceptual center of the modern nation-state. That territoriality is defined legally, as in the concept of "jurisdiction." The notion of "rights" is thus implicated, including the rights of a people to a land but also "private" rights. Who are, in fact, the people? Who belongs and has rights and thus partakes in shaping the civic order? This is an issue that concerned the Puritans and became, though not defined in those terms, an immigration issue.

Two observations can be made here: first, law came to be centered in territorial issues (in contrast to medieval practices, which rooted legal status in personal status groups); second, the legal structure is conceptualized as a manifestation of the sacred.[32] Even in later secular nationalisms, legal constructions were the means of embodying the national myth (in the sense of how the nation conceived of itself), most notably in the notion that the practice of self-government is expressive of the will of a people. Government was a self-conscious legal construction (think of the United States Constitution), and government was in the business of making laws.

The issue of "belonging" occupied the Puritans, the historical mirror (if with a different vocabulary) to later debates on immigration and citizenship. In the process of the community defining itself, questions of inclusion and exclusion became central. The covenant was by definition an attempt to assemble the godly and exclude the degenerate. Thus one covenant, of the community of Dedham, stated, "We shall by all means labor to keep off from us all such as are contrary minded, and receive only such unto us as may be probably of one heart with us . . . [this] for the edification of each other in the knowledge and faith of Lord Jesus."[33] So even if the Puritans contributed to America's self-conception as a nation of immigrants through their own Great Migration or as the source of national myths such as America as an asylum, they also illustrated that a "nation of immigrants" does not imply that it was a nation without boundaries. On the contrary, through the process of weaving together territorial, communal, and political identities into a singular, unitary, and bounded entity, the patrolling of social boundaries became more of a concern, not less.

The Puritans' concern with who did and did not belong was reinforced by their form of government. The state, as Perry Miller and Thomas Johnson write, was not an "umpire" arbitrating between different interests and different communities The state was instrumental in promoting and enforcing conformity to the covenant and maintaining a godly society of Puritan saints. It was a "dictatorship" of the "holy and regenerate." Those with other beliefs could stay out: if they came to New England, they should keep their opinions to themselves; if they publicized their opinions, they were exiled; and if they persisted in coming, they could be—as were four Quakers on Boston Common—hanged.[34]

The discussion of who to include or exclude reveals the nature of the "public." The notion of a civic or public sphere in the Puritan and republican context presupposed a common community. As is argued in the second half of this book, contemporary ideas of public space have turned this idea on its head (or feet, depending how one evaluates this development). The public arena, in this more classical approach, was one of engagement, not disengagement, a space of shared communion and identity, to be preserved and nurtured. The Puritans preserved that sense of themselves through the covenant; the community was to be governed by fundamental laws of Scripture. Reinforcing a legalistic form of government, ultimately magistrates and officers were accountable to those rules—an important source for the latter-day phrase "a nation of laws"—and the people who had committed themselves to God ultimately decided whether any action was in accordance with the covenant. Church elders were there to lead and preserve godliness, lest the people lose their way, carriers as they were of a Puritan orthodoxy and thus public conformity. (As one New England Calvinist wrote, "Let the elders publicly propound . . . let the people of faith give their assent to their elders' holy and lawful administration.") But all those bound by the covenant of a particular community, that is, the public itself, had to assent to new members and to expel heretics and undesirables.[35]

To preserve that public self, the regulation of newcomers was centrally important. The New England Puritan communities demanded faith and godliness, rather than wealth or property, as prerequisites for citizenship—which sounded odd to noblemen who considered immigrating to New England. If such fellowship were given "to men not according to their godliness," the Puritans stated, "but according to their wealth," their communities as a whole would become "no better than worldly men." Worldly men may then become a significant part of the magistrates, and that could "turn the edge of all authority and laws against the church and the members thereof, the maintenance of whose peace is the chief end which God aimed at in the institution of the Magistracy."[36] This argument anticipates, surely, and even informs, contemporary contentions about the importance of determining the "character" of potential immigrants and citizens.

These conceptions of civic society and politics also contributed to emerging ideas of popular sovereignty. For a political community to be a community, sovereignty has to be located somewhere. Sovereignty is what distinguishes the state, the body politic, from other forms of association. A supreme authority, theorists of sovereignty held, is necessary if there is to be a community at all. A world of purely private rights does not sanction the protection of general interests. Other associations, such as churches, trade unions, or businesses, have a "will," but their wills are subject to judicial appeal. Such bodies lack, in other words, the sovereignty of the state. The state (or, in the case of the Puritans, state *and* church, as they were mutually reinforcing) is the common bond, the one obligatory form of association shared through a singular citizenship.[37]

More broadly, the Puritans contributed to (and, many would argue, played a critical role in) the forming of modern politics.[38] The emergence of the citizen—today, the secularized saint—signaled, Michael Walzer observed, the integration and participation of private men in the political order. Politics was a "conscientious labor"; politics was the activity of constructing and defining the very character of social and political relations. Even more striking, however, was how the public character of the state insinuated the dialectic of state and society, of a public sphere and private association, into the language of modern political history. In constructing and shaping the political order, citizens made public presentations of their demands. And to make those demands groups were formed; civil associations were born, in principle united voluntarily and based on ideas, not blood. The rise of the state witnessed the rapid spawning of new organizations, political parties, and organizational initiative. Political activity demanded organization and programmatic expression of purposes.[39]

It is important to stress, however, that civic ideals and practices were not hostile to social rank as such or any other forms of inherited or ascriptive status. Ascriptive status, in contrast to achieved status, suggests that rank is a function of family of birth (as in the way aristocratic status was generally determined) or some other objective condition. Puritan writings do not appear to display ascriptive biases, except in a qualified sense with regard to women and American Indians. Ascribed status based on race, sex, and property does, however, obviously come to play a defining role in American history. Arguments that flow out of the writings of Alexis de Tocqueville, who claimed that American identity was in essence "invented," free of social rank of any kind, and that this had its roots in the middle-class and Protestant origins of the country, have in recent years come under fierce attack.[40] But apparently lost in this debate is how ascriptive status can be easily woven into civic practices, for civic politics inherently involves defining the character of the body politic, including who belongs and who does not. Such self-definition does not necessarily involve, by any means, exclusion based on inherited or ascribed characteristics. But it can. Thus racial exclusion, for example, can come to be part of civic practice in ways that would not be viewed as incongruous. The Afrikaners, for example, a Calvinist people like the Puritans, viewed themselves as republicans; for them their civic identity, until the collapse of apartheid, was one that associated Christianity, "whiteness," and the *volk* (their sense as an "organic" people) as singularly related. The exclusion and oppression of blacks followed from these beliefs and was internally viewed as morally quite unproblematic. Civic society was limited to whites.[41] If the civic or national self is rooted in blood descent, in a *Volksgemeinschaft*, exclusionary practices follow.

Puritans in the Wilderness

The Puritan poet Michael Wigglesworth composed his most memorable poem, *God's Controversy with New-England*, in 1662. It is a lengthy poem, but certain stanzas captured the attention of scholars discussing Puritans and the wilderness. Wigglesworth draws us into the dark foreboding that these early immigrants felt as they approached the coast of New England:

Beyond the great Atlantick flood
There is a region vast,
A country where no English foot
In former ages past:
A waste and howling wilderness,
Where none inhabited
But hellish fiends, and brutish men
That Devils worshipped.

This region was in darkness plac't
Far off from heavens light,
Amidst the shaddows of grim death
And of eternal night.
For the Sun of righteousness
Had never made to shine
The light of his sweet countenance,
And grace which is divine

In writings on Americans and their environment, Wigglesworth's evocative and dark description—"A waste and howling wilderness"—is said to capture the Puritanical view of the wilderness. The historian of the environment Roderick Nash pointed to Wigglesworth and this brooding vision in his study. Nash wrote of the Puritans' Manichaean imagery of the forces of light doing battle against the forces of darkness such that the wilderness came to represent darkness and satanic emanations. Puritans believed "God was on their side to destroy the wilderness." The driving desire, Nash wrote of the Puritans, was to stake out a garden in the surrounding darkness, making their sanctuary paradoxically their enemy at the same time.[42] Leo Marx in his book suggested that the "hideous wilderness" was characteristic of Puritan thinking. He quotes William Bradford's recollection on arriving in the New World as looking across the water and seeing "a hidious and desolate wildernes, full of wild beasts and willd men . . . the whole countrie, full of woods and thickets, represented a wild and savage heiw."[43]

Yet, if one continues on to read *God's Controversy with New-England,* following the above-mentioned stanzas, Wigglesworth's poem takes on a different hue:

Until the time drew nigh wherein
The glorious Lord of hostes
Was pleasd to lead his armies forth
Into those forrein coastes.
At whose approach the darkness sad
Soon vanished away,
And all the shadows of the night
Were turnd to lightsome day . . .

O happiest of dayes wherein
The blind received sight,
And those that had no eyes before
Were made to see the light!
The wilderness hereat rejoc't,
The woods for joy did sing,
The vallys and the little hills
Thy praises ecchoing.

Here was the hiding place, which thou,
Jehovah, didst provide
For thy redeemed ones, and where
Thou didst thy jewels hide
In per'lous times, and saddest dayes
Of sack-cloth and of blood,
When th' overflowing scrouge did pass
Through Europe, like a flood.

The wilderness rejoiced, and the woods for joy did sing! The Puritans had, then, a much more nuanced, complex relationship with the wilderness than is often portrayed.

The wilderness was fecundity, promise, seeding, nature untamed, and a representation of the sacred. The wilderness reflected humanity's moral state. The wilderness was life, progress, and movement. The wilderness was Sinai *after* the escape from the fleshpots and bondage of Egypt, on the way to the Promised Land.[44] The wilderness was genesis, literally and metaphorically. The wilderness was a refuge, a virgin soil unsullied by Europe's corruption. The wilderness was, simultaneously, a wasteland, a dark brooding place of demons. The wilderness was a source of life and a source of pestilence. God's hand was present in every hillock, every sweeping, exhilarating landscape, in the fresh, lush green fertile promise of America. But his swift retribution for human failings was expressed in each thunderclap, in droughts, floods, and infestations of locusts and worms. The wilderness was a source of awe and all that awe can invoke—excitement and fear. The moral complexity of humankind was reflected in the symbolic complexity of the wilderness for Christianity as a whole and especially, perhaps, for the Puritans.[45]

The wilderness was a *promise;* it was "here" that the world would be redeemed, where the forces of darkness would be vanquished. Thus the wilderness could evoke soaring thoughts and desires in its promise, yet it could also elicit a sense of the arduous road ahead. It was here that the apocalypse would take place. It was here where God had appointed his people to do battle to bring on the Kingdom of Heaven on Earth. It was here where the Reformation would be realized—but that task of fulfilling the redemptive promise was still before them. It was also here, however, that the demonic forces—internal and external—would have to be faced. The covenant brought them to the Promised Land, but the covenant also suggested an agreement, an agreement with God, which by definition still had to be fulfilled. The Puritans were at home in the New World but not entirely at peace with themselves: the Lord's work still had to be completed. In a sense, the Great Migration did not end in Massachusetts Bay, but the journey—spiritual, political, and historical—had just begun.[46]

The wilderness was to be cultivated and transformed, capturing the dynamic, "progressive," historically driven sense of movement to the eschaton. "The whole earth is the Lords garden," wrote Winthrop, "and he hath giuen it to the sonnes of men with a gen[eral] Commission: Increase and multiplie, and replenish the earth and subdue it . . . that man might enioy the fruits of the earth, and God might haue his due glory from the creature." Elsewhere he wrote of the sons of Adam tilling and improving the land.[47] But this did not mean the wilderness was to be "controlled" or "destroyed," as some have suggested and often popularly believed. The repeated references in Puritan writings to the metaphor of "planting" and references to Genesis where "the Lord God planted a garden eastward of Eden . . . And out of the ground the Lord God made to grow every tree" captured a sense of being infused with and *harnessing* nature. The cultivated wilderness suggested fecundity, growth, progress, increase, calling, and labor. The land, the vegetation, and the soil were, in other words, part of the very soul of Puritanism and its ethic, contributing to, and anticipating, a theme in American national identity and, indeed, in all national movements.

In this light, it is telling indeed that Puritans condemned fashions in gardening which sought to shape trees and shrubs and to form geometrical gardens. Cultivating the wilderness did not mean "controlling" it; cutting and shaping implied a certain static quality detracting from the fecundity and the promise of growth and progress the Puritans saw in the wilderness.[48]

The ambiguous nature of wilderness—situated, as it were, between a Genesis-like chaos and order—did generate conflicting responses. The Puritan arriving in New England was like, in this limited regard, the pilgrim who finally reaches Jerusalem. The pilgrim finds a city redolent with sacred meanings and symbols but also a city concerned with the mundane tasks of everyday life and characterized by many petty corruptions. The pilgrim finds Jerusalem, so to speak, suspended between the City of Man and the City of God, engendering a certain ambivalence, a state of liminality. This was the city in which the Messiah would arrive, but the Second Coming was not yet here. Thus the New World environment engendered remarkably divergent responses. The Reverend Francis Higginson, a spiritual leader among first-generation Puritans sent to Salem, commented: "The Temper of the Aire of New-England is one speciall thing that commends this place." Experience, he continued, "doth manifest that there is hardly a more healthful place to be found in the World that agreeth with our English Bodyes. Many that haue beene weake and sickly in old England, by comming hither hane beene thouroughly healed and growne healthful and strong."[49] William Bradford, on the other hand, wrote, that the "chang of aire, diate, and drinking of water would infecte [the Pilgrims'] bodies with sore sickness, and greevous deseases."[50]

The nuanced, sometimes contradictory, approach of the Puritans to the wilderness was also a product of their seeking redemption *in* this world while never being *of* this world. Being monotheists who believed in a transcendent God, rather than in a pantheistic god who literally inhabited nature, they held a certain suspicion and often certain knowledge that the material world was corrupt. On the other hand, unlike their other-worldly Protestant brethren who believed that this world was incorrigibly corrupt and that thus one sought to escape it mystically, the Puritans thought the world could be redeemed through political and military action, and they were the ones chosen to do it. The Puritans believed in, Perry Miller and Thomas Johnson observe, the "fallibility of material existence and the infallibility of the spiritual, the necessity for living in a world of time and space according to the laws of that time and that [space], [while] never once forgetting that the world will pass [and] be resolved back into nothingness, [and] that reality and permanence belong to things not as they appear to the eye but to the mind." More precisely capturing Puritanism's paradoxical relationship with this world, Max Weber wrote in one of many memorable passages in *The Protestant Ethic:* "[Catholic] asceticism, at first fleeing from the world into solitude, had already ruled the world which it had renounced from the monastery and through the Church. But it had, on the whole, left the naturally spontaneous character of daily life in the world untouched. Now [Protestant ascetics] strode into the market-place of life, slammed the door of the monastery behind [them], and undertook to penetrate just that daily routine of life with . . . methodicalness, to fashion . . . a [pious] life in the world, but neither of nor for this world."[51]

Ultimately, perhaps, the Puritans' relationship to the "wilderness," in the broadest sense, lay in their subtle, if complicated, conjunction of time and space, or in other words, in their sense of place. Demarcating a place inherently involves defining an area as enclosed, as discrete, as having an "interior," and, in effect, designating other spaces as "outside." This is, in contrast, different from time. Time, in its linear and chronological sense, is continuous, successive, and accumulative. Space is distinctive, time is incorporative.[52] This dialectic between space and time, distinctiveness and incorporation, becomes even more marked in later American history, after independence and after Puritans themselves disappear as a definable constituency (in the early nineteenth century). As discussed later, the United States for its first hundred years or so was never actually clear as to its territorial confines or limits, its borders, a point of significance in of itself as to America's place in the world. (Similarly, related to this phenomenon, the United States was not sure, until after the Civil War, it was one nation.) "The nation," Daniel Boorstin wrote, "was beginning not at one time or place, but again and again."[53] This quality of beginning again and again also ties to notions of history and America's place in it: dynamic, "progressive," expansionist, and "incorporative,"—these elements necessarily came together. Like perhaps no other nation, America's historical role has made its "place," in the physical as well as in the moral sense, problematic: its self-described historical place as the carrier of universal values made borders and boundaries constraining, even impediments. History was accumulative; territory, again, was enclosed. This dialectic of time and space becomes an ongoing, implicit theme in the American story.

The Wilderness as Resource and as Basis of Scientific Knowledge

If a painting is a fusion of a subjective sense, of the artist or of a public sensibility, with a material form, so the landscape can be conceived of as a painting, rendered in different ways in different social contexts. For the Puritan, the patterns, uniformities, ebbs, and flows of the landscape and of the wilderness reflected the beauty of God, and insofar as there were "laws" of nature, they surely reflected the guiding hand of God as well. The wilderness was also a divinely provided economic resource. Winthrop talked of "Godes blessing vpon the wisdome and industry of man . . . and whatsoeuer we stand in need of is treasured vp in the earth by the Creator, and is to be fetched thence by the sweatt of our browes."[54] Just as God plants his people, so the people through cultivation plant themselves: through cultivation, the people come to have a moral as well as an economic relationship with the land. The wilderness as the object of scientific study, or as a resource, and the wilderness as the sacred representation of God were not viewed by the Puritans as a contradiction. On the contrary, science and God, industry and land, were complementary.

The landscape is "a mappe and shaddow of the spirituall estate of the soules of men," said Cotton; or as Jonathan Edwards, one of the last of the great New England Calvinist theologians, exclaimed, "When we behold the fragrant rose and lily, we see His love and purity." Or, in the words of Johann Alsted, a seventeenth-century German thinker who influenced the Puritans, "A genuine reading of the book of nature is an ascension to the mind of God."[55] The landscape, the firmament, was the ineffable expressed in a material form. The "visible world" was the way through which the divine presence communicated with the believer. Thus nature played all the more an important role for the Puritan, who steadfastly rejected the other-worldly mysticism of Protestant sects that sought a mystical union with God. As a consequence, the study and understanding of the laws and workings of nature are not only acceptable but even a duty as a requisite part of Christian knowledge and of knowing God. God worked within nature, not by violating it or its laws, to the point where it was believed that even God had to obey the natural laws he created. The ordered, patterned forms of nature reflected God's beneficence and beauty and his accessibility through reason. In this thinking, we see the beginnings of the

Enlightenment in America and the seedbed of thinking that would influence, among others, Paine and Jefferson.[56] We also see the merging of aesthetic beauty and scientific form.

The worldliness (in the sense of being oriented to reforming, rather than escaping, this world) of the Puritans produces then, paradoxically, a faith in the laws of nature. Everything was explainable, and nothing was accidental. In this we see the precursor, or the historical mirror, of Newtonian science and the "nonrational" roots of the rational that Max Weber would later point to about the modern world.[57] It is taken on faith, first by the Puritan saint and later by the modern rationalist, that everything can ultimately be explained.

American Indians and the Struggle to Define the Land

In William Bradford's history of Plymouth Plantation, his descriptions of the native peoples express a sense of Christian superiority and preeminence on the one hand and fear on the other. The Pilgrims, when considering where to immigrate, turned their thoughts, Bradford wrote, to "those vast and unpeopled countries of America, which are fru[i]tfull, and fitt for habitation; being devoyd of all civill inhabitants, wher ther are only [sa]vage and brutish men, which range up and downe, litle otherwise then the wild beasts of the same." He continued, describing the Separatists' fears, "[The Pilgrims would be] in continuall danger of the [sa]vage people, who are cruell, barbarous, and most trecherous, being most furious in their rage, and merciles wher they overcome; not being contente only to kill, and take away life, but delight to tormente men in the most bloodie manner that may be; flaying some alive with the shells of fishes, cutting of the members and joynts of others by peesmeale, and broiling on the coles, eate the collops of their flesh in their sight whilst they live; with other cruelties [too] horrible to be related."[58] This sort of account has come to typify, in the minds of many, the Puritan as well as Pilgrim approach to the Native Americans, an arrogant self-assurance in a brutal one-sided struggle over the land. The historian Sacvan Bercovitch, for example, states, "What [Puritan documents] tell us, in effect, is that there are two parties in the new world, God's and the Devil's; and that God's party is white, Puritan, and trusted with a world-redeeming errand, while Satan's party is dark-skinned, heathen, and doomed."[59] Although clearly the story of American Indians in the face of European settlement and expansion is almost too "horrible to be related" and brutal, this perspective of an essentially racist "land-grab" hides as much as it reveals with regard to the Puritans, certainly in the early years of the settlement.

American Indians were not defined by the Puritans, at least initially, in terms of racial categories. They had a "moral" status.[60] This moral status was not generally in the Indians' favor, to say the least. However, to the extent that Native Americans were viewed as having evil, barbarous inclinations and as being possessed by the devil, such characteristics were not thought of as somehow racially inherent. If the Indians were working at the behest of Satan, it also meant that they could be "saved." This moral dimension is similarly tied to the issue of land. Land is a resource, and land is a form of power, certainly, but resources and power are mediated through certain cultural and moral filters and understandings. Those filters fundamentally affect the forms of political and social relations which take place on the land. The Puritans were making a moral claim to the land, and the presence of indigenous peoples—"savages" though they may be—*was* viewed by Puritan divines as a fundamental moral problem that had to be addressed. In a poignant passage towards the end of his sermon *God's Promise to his Plantations,* John Cotton beseeched his flock: "OFFEND NOT THE POOR NATIVES; but as you partake in their *Land* so make them partakers of your *precious faith:* as you reap their *Temporals,* so feed them your *Spirituals.* Win them to the love of Christ, for whom Christ died. They never yet refused the Gospel, and therefore more hope they will now receive it. Who knoweth whither God have reared this whole Plantation for such an end?"[61] How revealing the sentence "Who knoweth whither God have reared this whole Plantation for such an end?" The Puritans had a sacred mission, imbued with a deep sense of historical purpose, to reform this world. Reforming the heathens (who have "never yet refused the Gospel") through conversion was an extension of that mission.

The Puritan approach to the American Indians provides us, in fact, with a prism that brings into focus the interdependent linkage between territoriality, civic politics, and, by implication, nationhood. In justifying possible displacement of the Indians—and this was considered a serious moral issue by the Puritans, discussed repeatedly—one central theme addresses how a moral (and by extension, social and political) claim to a land can be made. The Puritans conclude that a people that are nomadic, or have no fixed, "bounded," and enclosed territories with exclusive, sovereign control, cannot make a moral or legal claim to that land. Native American groups that were nomadic, or had amorphous boundaries, or shared lands with other Native American tribes, failed on this test. This was not simply a convenient stratagem (although it could and was later used that way) but went to the essence of notions of the "civic" (or what other Europeans would refer to as "civilized") and, the opposite, "savagery" and "barbarism." It is through the civic sphere that people are defined as a national and political entity; the civic identity is a legal construct, not one of kinship as such. The people's presence, and through them the presence of God, are demarcated by their legal and territorial jurisdiction. As noted earlier, territory becomes the locus of bounded, spatial communion, the source of historical memories and expectations. The land is their "home," which defines their "place" in the world. In this context, the American Indians, with the importance of kinship and with radically different ideas of territorial association, be it nomadic or tribal, appeared to the Puritans as rootless, and in some savage state. It is not legitimate, the Puritan divines stated, to settle where the native peoples actually reside or have cultivated the land. But the broader associations with the land, such as those of nomadic bands, are not recognized. The Puritans could hold these views even though they recognized that "nature" as such held an enormous significance, not only economically but in terms of religion and mythology, for Native Americans.[62]

Winthrop, writing in 1629, responds to different objections to the proposed migration to America, one being, "But what warrant have we to take the land, which is and hath been of long tyme possessed of others the sons of Adam?" Thus the Indians are recognized as fellow sons and daughters of Adam, not, in some biological sense, as different beings. Winthrop responds:

> That which is common to all is proper to none. This savage people ruleth over many lands without ti[t]le or property; for they inclose no land . . . [They] remove their dwellings as they have occasion, or as they can prevail against their neighbours. And why may not christians have liberty to go and dwell amongst them in their waste lands and woods (leaving them such places as they have manured for their corne) as lawfully as Abraham did among the Sodomites? For God hath given to the sons of men a twofold right to the earth; there is a naturall right and a civil right. The

> first right was naturall when men held the earth in common . . . Then, as men and cattell increased, they appropriated some parcells of ground by enclosing . . . and this in tyme got them a civil right.[63]

Similarly, Cotton addresses at some length arguments justifying settlement where native peoples may be. God, he said, makes room for "a People" in three ways: (1) through lawful war on the heathens—but this is exceptional, demanding a "special commission" from God; (2) through purchase from, or at the "courtesie" of, the "native people"; and, finally, (3) when a land, though not altogether devoid of inhabitants, is clear in places where the new migrants are residing. "Where there is a vacant place," Cotton stated, "there is liberty for the Son of *Adam* and *Noah* to come and inhabit, though they neither buy it, nor ask their leaves." Interestingly, Cotton does not suggest, again drawing on the story of Abraham as support, that a calling from God be used to justify the settling of a land, for, he said, "that would have been deemed frivolous amongst the Heathen." Instead, it is "a Principle in Nature, That in a vacant Soyle he that taketh possession of it, and bestoweth culture and husbandry upon it, his Right it is."[64] These arguments were a repeated refrain in the New England experience. (Winthrop also noted, for good measure, that there was more than enough land for "them and us" and that "God hath consumed the natives with a miraculous plague, whereby the greater part of the country is left voide of inhabitants." Winthrop also adds, in less macabre fashion, that settlement is also legitimate when "we shall come in with good leave of the natives.")[65]

From the perspective of Native Americans, being "saved" or being forcibly removed still threatened the dissolution of their culture. But the Puritans, and even more so, the Quakers, were in practice, Charles Segal and David Stineback write, less prone to forcibly acquiring Native American lands than their fellow compatriots along the mid-Atlantic coast. It was this latter group who had no interest in conversion or who did not have an interdependent economic relationship with the Indians, who were most likely to revert to warfare over land. However, the Puritans' relative restraint diminished when fighting American Indians allied with the French, for then they were viewed as instruments of the Catholic heresy, not as savages who had "not yet refused the Gospel." And as American Indians did not convert to the extent Puritans had hoped, attitudes also hardened, to the point that it was even felt that the Indians were "constitutionally unable" to become good Christians. The dissipating influence of the Puritans as the American Revolution approached lessened the theological role of native peoples (with mixed results, at best, for the Indians), but certain Puritan concepts remained. Nineteenth-century Americans still believed that shared territories, nomadic paths, and amorphous boundaries were not a basis for territorial claims. Instead Native Americans, if not outright killed, were forcibly moved onto bounded, defined, and semisovereign reservations. It was in such bounded territorial zones that remnants of the native peoples and their cultures could be, or were allowed to be, sustained.[66]

THE MINISTER William Hubbard, in a sermon given in 1676, said that it was "Order that gave Beauty to this goodly fabrick of the world, which was before but a confused Chaos, without form and void."[67] Puritanism linked the natural order to a political, moral, and civic order. From this sentiment, it would be but a short leap to references to laws of nature and "Nature's God," in the Declaration of Independence and the Constitution.

Surveying the Landscape: Place and Identity in the Early Republic

TWO

O beautiful for spacious skies,
For amber waves of grain,
For purple mountain majesties
Above the fruited plain!
America! America!
God shed his grace on thee
And crown thy good with brotherhood
From sea to shining sea!

—KATHARINE LEE BATES

AMERICA. The very word conjures up desert landscapes, broken by temple-like rock formations and perhaps the Marlboro cowboy riding by on his horse. Or the plains with their "amber waves of grain." Geological marvels are pictured: the Grand Canyon, the Yellowstone geysers, the Dakota Badlands, or Niagara Falls. The human presence is also felt in places like Mount Rushmore, Gettysburg, and the Alamo. Americans—and indeed people across the world—tend to imagine America in terms of its landscapes. Of course, there is that *other* America: Disneyland, New York, Hollywood, Las Vegas, urban blight, rural poverty, highways, crime, and violence. But if the allusion is to the *soul* of the country—that is, the myths, symbols, images, and ideas that bind Americans and makes them think of themselves as one people—it is the landscapes, the land, and places that are most often evoked. Community is thought to be local and neighborly, and democracy is civic participation in townships, villages, and local churches. When the reality fails to support this picture, this moral geography, something is wrong with America—it is in civic and moral decline.

Songs and speeches that stir emotions and patriotic attachments similarly evoke distinctive landscapes and places, the "beautiful spacious skies." What these landscapes reflect, however, is the early, mostly nineteenth-century Republic. "Thinking ourselves across space," Edward Ayers and Peter Onuf observe, "we think ourselves backward in time, imaginatively returning in an idealized past." American geography, they continue, represents American history. Americans have "spatialized time and historicized space."[1] Nineteenth-century America is still with us, though the conversation about people and place has taken new turns. For the citizens of the pre–Civil War Republic, locating "place" and identity was, quite likely, *the* debate. They inherited from the Puritans a firm belief in the inextricable relationship of the people to clearly demarcated places; the question was where to mark those boundaries and, by extension, who were, in fact, the "people."

In the United States after 1776 we see not only a continuation of the Puritan sacralization of the land but a veritable communion with it (as expressed in the art of the Hudson River School or the philosophy of Thoreau); and the association of the land with republicanism and nationhood is broadly celebrated in the "agrarian myth." As with the Puritans but in a more elaborated form, the land is celebrated as the source of economic growth and scientific knowledge and as the expression of God's presence. It was to be harnessed in America's march forward in history. So there is the same seamless weave of land, politics, and peoplehood which can be observed with the Puritans. But there are significant shifts, as well. The United States covered a larger swath of territory than New England and was made up by a multiplicity of states. In addition, in contrast to the almost monolithic Puritans (certainly in terms of their theology and politics), the United States was more multidenominational, characterized by a diverse array of sects and churches that represented a distinctly different religious and social outlook: a more pluralistic view of religion, at least some readiness for separation of church and state, and an acceptance of individual prerogatives and mobility in belief, church membership, political allegiances, and, to some extent, social practices—at least for whites.

In contrast to state churches, such as the Puritans or the Church of England, denominational religious organization tended to be decentralized, voluntary, and formed locally.[2] Such denominations, in a time when the broad American public had some form of religious affiliation, displayed varying degrees of suspicion about government power. (Tom Paine, himself of Quaker origin, began his tract *Common Sense* by referring to government's "wickedness," a sentiment that resonated and was seized upon in the most widely distributed propaganda pamphlet of the American Revolution.) Starting with thirteen states, numerous sects (albeit overwhelmingly Protestant), suspicion of government, and a growing individualism, people generally "placed" and identified themselves in terms of states or smaller geographic units. The federal Republic was even created, in 1789, by states (in contrast, by 1861, the eve of the Civil War, most states were born into the world by the federal government), and it was not until after the Civil War that the United States was referred to in the singular, rather than the plural (as in "the United States are a republic").[3] The Constitution of 1789 would create a *union* but not, at first, a singular *nation*.[4]

The association between people and place was shaped and negotiated in certain iconic writings and enactments in the antebellum United States. These include the essential arguments or stories of Thomas Jefferson, Thomas Paine, and J. Hector St. John Crèvecoeur on the special association of the land, people, and republican politics in the United States. Such ideas were institutionalized through "definitive events or moments" that played a framing role in channeling future developments. One document in particular, the Northwest Ordinance of 1787 (discussed at length later), which many believed to be of "high, constitutional importance," played such a central role in mapping American expansion. In the Jeffersonian tradition, which in critical (often edited) respects became the touchstone of antebellum American politcs, republicanism stressed individual freedom and majority rule, to be realized in small geographic units, such as the township. Even the justification of the Constitution and stronger federal government was primarily on instrumental grounds, and it was largely presumed that in a federal framework republicanism worked best on the "local" level. In the Northwest Ordinance of 1787, this "moral geography" of small republics was instituted, and the federal government was bequeathed the critical role of overseeing expansion in the West. However, in the process of expansion, the growing role of federal government, and its looming legal presence, we see the gradual and uncertain centralization of place—and, in turn, nationhood. Concomitantly, the two strands of Jeffersonianism—individual liberty, mythologized in the independent farmer, and local or state sovereignty—clash in painful and irreconcilable conflict over the slavery issue.[5]

Slavery and democracy, it became clear, had distinct "spatializations," distinct notions of place, space, and time. Slavery was associated with large plantations, pastoral imagery, an almost feudal leisure class, and a sense of timelessness. (Such a picture also demanded, of course, people or peoples who had been torn from their land and stripped of any moral association or claim to the soil they worked.) Republicanism rested in the virtue of the farmer, homesteads, relatively small farms, and restless, hardworking yeomen with a vision of a better future, of ongoing progress.[6] The farmer cultivated the soil and in doing so cultivated—or *civil*ized—him- or herself. The farmer's relation to the land was through legal title and through work. The slave owner belonged to an "aristocracy" based on blood. Although this imagery took on mythic proportions (in the sense of touchstones of American identity as well as "idealized" conceptions of the national self), the myths were sown into the land—in the way it was mapped, townships zoned, crops planted or not planted, and so on. Such "geographical lines" coincided with a "marked principle, moral and political."[7] These different landscapes reflected different moral geographies, as well as moral economies, and as such came into irreconcilable conflict. Thus, northern workers could feel threatened by slavery while not being sympathetic to African Americans: the homestead could symbolize certain values of independence and civic freedom which the plantation could not, even to racist northerners. It became a struggle against slave states but not *necessarily* a struggle for inclusion of black Americans.

Beyond the slavery issue, from Independence in 1776, the Northwest Ordinance in 1787, and the Constitution in 1789, through to the outbreak of the "War of the States" in 1861, America is mapped—literally, of course, but also "constitutionally." In the absence of kinship principles to define a society, geographic demarcations become constitutive of the nation. To define an identity, writes Thongchai Winichakul, requires the drawing of lines, of ins and outs, which is inherently a spatial process; the construction of national identity is, similarly, a spatial process. That "space," in nation-states, is the soil, the territory—territory in this sense constitutes the nation. It is the most concrete, "real" feature of the modern nation. It is the material form, what Winichakul calls the nation's "geo-body."[8]

A revealing aspect of the cartography of modern states is its precision and its unambiguous quality. This precision is critical because the act of mapping is so important in defining the body politic and who is inside and who is outside (which is also critical in defining the quality of movement, such as "immigration"). In contrast, the border zones of other polities and civilizations were often ambiguous or staggered in different ways, such as Confucian China's Middle Kingdom, Siam, medieval Europe, and the Middle East and Africa before European colonization. One pattern that can be discerned, tied to the larger theme in this book, is that where kinship principles predominate as the basis of rule (such as kingship) and where the sentiment of a "corporate" people is absent or marginal, borders are less likely to be demarcated carefully and precisely.

The People and the Land

In the popular imagination, the founding of the United States is often perceived as the heroic establishment of democracy in the face of the tyrannical British monarch in a time when, indeed, Europe as a whole was dominated by autocrats. Even the texts of histories are variations of this theme, though sometimes challenging the heroic or democratic character of the enterprise; for example, some historians, such as those associated with Charles Beard, have suggested that the founders were driven by venal economic goals or, more recently, that the struggle was not a democratic or republican revolution at all but one that served only the interests of the white, male, and propertied.[9]

But in all these debates, arguments, and celebrations over the birth and subsequent growth of the Republic, one particular yet critical characteristic is largely overlooked. Europe *was* the counterpoint of reference to the revolutionaries and in the early Republic. Dark despotic Europe, the Old World, was the foil to this vibrant, new nation that would be a beacon of light, a City on the Hill, in a world of despots. And this imagery is still with us, democratic America versus despotic Europe. But what is largely forgotten today, at least in the popular memory, is that the Americans then were comparing not just political systems but different countries, in the dual sense of the term—the landscape and the nation. Americans were comparing their *land* with Europe, describing the moral qualities of their land and its special relationship with the American people. European societies, in contrast, were characterized by overbearing "tradition," a stress on kinship and blood descent of royal families and aristocrats, and a land in decay from overpopulation and crumbling buildings. (Some Europeans even today, in a ongoing pique of jealousy perhaps, like to turn this argument on its head. Americans, they suggest, "lack culture," meaning high culture, and are ignorant and unsophisticated.) The Americans also compared their civic and republican political culture with the European monarchical alternative.[10]

This remarkable emphasis on the land, the country, was viewed as root and branch of this new republican and national experiment, as the soil that nourished both their civic politics and self-concept. The reference to their *natural* rights as individuals and as a people was understood quite literally—their rights were vested in the land and in "nature." Thus the relationship between the land, the people, and their civic polity was, on the one hand, "unproblematic" and, on the other, to be celebrated, defended, fought for, and, of course, "lorded" over the British and the Europeans. The binding tie to the soil of the people and their brand of politics, and the contrasts with the decrepit and degenerate Europeans and their traditions, are pervasive, characterizing both political documents and popular culture. The distinctions, then, went beyond simply republicanism versus monarchy; they were distinctions between different *places,* where ideas of history and geography intersected in very different ways.[11]

July 4, 1776, was a moment when, it was felt among the patriots, heaven touched earth. How telling, then, that the very first sentence of the Declaration of Independence reads, "When, in the course of human events, it becomes necessary for one people to dissolve the political bonds which have connected them with another, and to assume among the powers of the earth, the separate and equal station to *which the Laws of Nature and of Nature's God entitle them,* a decent respect to the opinions of mankind requires that they should declare the causes which impel them to the separation." America—its people and its republican politics—was act of nature. "Nature," of course, can be construed in different ways, but here and clearly for Thomas Jefferson, the author of the Declaration, nature was not abstract; it was the soil, the vegetation, the very life of this earth.[12] The flattering contrast (that is, for Americans) with Europe was expressed in popular literature. Henry Van Dyke, writer, poet, and United States ambassador to the Netherlands, expressed it thus:

> 'Tis Fine to see the Old World, and travel up and down
> Among the famous palaces and cities of renown,
> To admire the crumbly castles and the statues of kings,—
> But now I think I have had enough of antiquated things.
>
> So it's home again, and home again, America for me!
> My heart is turning home again, and there I long to be
> In the land of youth and freedom beyond the ocean bars,
> Where the air is full of sunlight and the flag is full of stars.[13]

This celebration of America, and of the symbiotic relationship of people and land, is expressed at the beginning of *The Federalist.* That document consisted of eighty-five short essays under the pseudonym Publius (authors Alexander Hamilton, James Madison, and John Jay) which serially appeared in New York newspapers in 1787 and 1788 as polemics in support of ratification of the Constitution. Not only is it recognized as America's preeminent contribution to political theory, but it is also considered a foundational legal source, along with the Constitution and the Declaration of Independence. In *Federalist* No. 2, "Concerning Dangers from Foreign Force and Influence," John Jay celebrated and promoted America, the land and the people. He first suggested that geography is destiny, its topographical features naturally, so to speak, demarcating the contours of the American people. In a passage worth quoting at length Jay wrote:

> It has often given me pleasure to observe that independent America was not composed of detached and distant territories, but that one connected, fertile, widespreading country was the portion of our western sons of liberty. Providence has in a particular manner blessed it with a variety of soils and productions, and watered it with innumerable streams, for the delight and accommodation of its inhabitants. A succession of navigable waters forms a kind of chain round its borders, as if to bind it together; while the most noble rivers in the world, running at convenient distances, present them with highways for the easy communica-

> tion of friendly aids, and the mutual transportation and exchange of their various commodities.

He then describes the sacred tie between the people, their land, and their political independence, clearly stating that the primordial unit—the ultimate reference, the ultimate point of vitality—was not the individual states or townships but the Union itself. And in a thread that goes back to the Puritans, the presence and guiding hand of God is felt:

> With equal pleasure I have as often taken notice that Providence has been pleased to give this one connected country to one united people—a people descended from the same ancestors, speaking the same language, professing the same religion, attached to the same principles of government, very similar in their manners and customs, and who, by their joint counsels, arms, and efforts, fighting side by side throughout a long and bloody war, have nobly established general liberty and independence.
>
> This country and this people seem to have been made for each other, and it appears as if it was the design of Providence, that an inheritance so proper and convenient for a band of brethren, united to each other by the strongest ties, should never be split into a number of unsocial, jealous, and alien sovereignties.
>
> Similar sentiments have hitherto prevailed among all orders and denominations of men among us. To all general purposes we have uniformly been one people each individual citizen everywhere enjoying the same national rights, privileges, and protection. As a nation we have made peace and war; as a nation we have vanquished our common enemies; as a nation we have formed alliances, and made treaties, and entered into various compacts and conventions with foreign states.

He then explains why, in the midst of revolution, the people turned to a loose confederation, rather than a relatively more centralized, federal government. Federal government would better reflect the purported unitary quality, in Jay's hopeful description, of the American people:

> A strong sense of the value and blessings of union induced the people, at a very early period, to institute a federal government to preserve and perpetuate it. They formed it almost as soon as they had a political existence;

> nay, at a time when their habitations were in flames, when many of their citizens were bleeding, and when the progress of hostility and desolation left little room for those calm and mature inquiries and reflections which must ever precede the formation of a wise and well balanced government for a free people. It is not to be wondered at, that a government instituted in times so inauspicious, should on experiment be found greatly deficient and inadequate to the purpose it was intended to answer.[14]

Jay's appeal, however, on the specifically *moral* association of an American people with their land *which transcended* all other geographic units and affiliations, above all states, faced an uphill battle. Other more "local" geographies, from farms to townships to states, would be equally, if not more, celebrated. Much of the argument for the federal government, as noted earlier, would be based on "practical" grounds, such as security and commerce, and its legitimization would often rest in its reinforcing of local and state authorities. The boundaries, in a moral sense, of "place" and "belonging," and what they implied about identity and nationhood and about political practices, would be the quintessential question facing the Republic through to the Civil War (the description of the war itself a Union presumption) and, in some respects, beyond. The intrinsic and seamless association between territory, people, and republicanism coupled with the difficulty of determining the exact jurisdiction in a (religiously) pluralistic and a multistate America is displayed in the thinking of Thomas Jefferson and Thomas Paine and in popular writing of the day.

"Nature's God" and the New Republic

On March 4, 1801, Jefferson presented his First Inaugural Address to his "friends and fellow-citizens." It was a lofty address, encompassing and binding together the republican vision, the salient role of law, the metaphor of a nation on a journey, the centrality of the different states, religious pluralism, and the sacredness of the land. "A rising nation," Jefferson declared,

> spread over a wide and fruitful land, traversing all the seas with the rich productions of their industry, engaged in commerce with nations who feel power and forget right, advancing rapidly to destinies beyond the reach of mortal eye[s]—when I contemplate these transcendent objects,

and see the honor, the happiness, and the hopes of this beloved country committed to the issue, and the auspices of this day, I shrink from the contemplation, and humble myself before the magnitude of the undertaking [before me] . . .

Kindly separated by nature and a wide ocean from the exterminating havoc of one quarter of the globe; too high-minded to endure the degradations of the others; possessing a chosen country, with room enough for our descendants to the thousandth and thousandth generation; entertaining a due sense of our equal right to the use of our own faculties, to the acquisitions of our own industry, to honor and confidence from our fellow citizens, resulting not from birth, but from our actions and their sense of them; enlightened by a benign religion, professed, indeed, and practiced in various forms, yet all of them inculcating honesty, truth, temperance, gratitude, and the love of man; acknowledging and adoring an overruling Providence, which by all its dispensations proves that it delights in the happiness of man here and his greater happiness hereafter; with all these blessings, what more is necessary to make us a happy and a prosperous people?[15]

Jefferson in this address, as he did in the Declaration of Independence, was drawing on a conception of republicanism which had become deeply rooted—almost literally. Republicanism embodied "natural rights"—in the words of the Declaration, "the separate and equal station to which the Laws of Nature and of Nature's God entitle them"—which existed independently of nations or time. But natural rights were also to be realized in a specific time and space; indeed, they were rooted in the specific soil of providentially blessed America. America did not so much escape history as overcome the backward-looking stress on tradition and kinship-based oligarchies.[16] The republican premises of changing this world, of self-determination, equality, and sovereignty, were, Jefferson recognized, not fully realized but could be imagined in the future. The United States was a chosen nation, a model of a new social order, and the redeemer nation of the world.[17] And yet this moral, redeeming quality was in tension with the very act of separation, anticipating an isolationist thread in American life. Republicanism could prosper, it was felt, if freed of all political ties with European despotism—America was "kindly separated by nature and a wide ocean from the exterminating havoc of one quarter of the globe," as Jefferson put it.[18] "The attempt to realize a better social order," wrote the historian Felix Gilbert, "presupposed a critical view of the values of the Old World and aroused a fear of ties which might spread the diseases of Europe to America."[19]

Certain critical premises of the Puritans were thus carried into the new Republic—God was "present" in nature, rights rested in nature, and, as such, the manifestation of rights, above all rights to collective self-determination, was "naturally" marked, that is, geographically demarcated and determined. The contrast between decrepit Europe and the natural grandeur of America, the desire to live in harmony with nature, and the symbolism of agrarianism and agrarian democracy suffused the social, cultural, and political life of America as a consequence. In a sense, these notions partly reproduced the Covenant or could be portrayed that way, as the historian David Noble argued: "The concept of a Biblical commonwealth was replaced in the eighteenth century by the Enlightenment's belief that the society of English colonies rested on natural principles and that the new republic that emerged from the American Revolution had a covenant with nature which freed it from the burdens of European history."[20] God had "planted" his people, in the imagery of John Cotton, and the soil nourished his people, and out grew a civic commonwealth—for all the dramatic social and political changes, that theme remained, *mutatis mutandis,* constant.[21]

In the writings that reflected a certain self-understanding of Americans of the early Republic—writings that consequently so resonated with the body politic that they in turn came to be the very touchstones of America's consciousness of itself—certain themes play out, implicitly or explicitly, the knitting together of land, republican politics, and the national self. Humans were a product of nature and as such had certain inherent "natural" rights. Those rights, as God created the natural and physical world, were imbued with a sacred presence. Humans were equal before God and, through nature, had an unmediated relationship with God.

If in the past kinship, through the monarchy and the aristocracy, was the primary source of position, status, and power, territory was now pointed to as the marker of community and identity. Rights inhered not in blood but in nature. Like the landscape around us, Thomas Paine argued, rights are "natural." God created the earth, and out of the earth God created humanity.[22] Furthermore,

Paine proclaims, the Mosaic account of creation (as well as every other account of creation), "whether taken as divine authority or merely historical," indicates that humankind was created as a "unity," meaning that no distinctions were made, aside from sex. This points to, for Paine, the fundamental equality within humanity. "Every child born into this world must be considered as deriving its existence from God," he wrote. The world, Paine continued "is as new to him as it was to the first man that existed, and his natural right in it is of the same kind."[23] Attacking the aristocracy, he said that "through all the vocabulary of Adam, there is no such animal as a duke or a count; neither can we connect any ideas with the words. Whether they mean strength or weakness, wisdom or folly, a child or a man, or the rider or the horse, is all equivocal. What respect then can be paid to that which describes nothing and which means nothing? Imagination has given figure and character to centaurs, satyrs, and down to all the fairy tribe, but titles baffle even the powers of fancy."[24]

Monarchy and rule and status by hereditary paths were a violation of nature and, by extension, of God. And such a violation could only have come through oppression. "Mankind being originally equals in the order of creation, the equality could only be destroyed by some subsequent circumstance," Paine wrote in *Common Sense.*[25] And elsewhere, in the *Rights of Man,* he notes that there was "a time when kings disposed of their crowns by will upon their deathbeds and consigned the people, like beasts of the field, to whatever successor they appointed."[26]

The American Revolution, by sweeping away monarchy and aristocracy, brought people back to nature, to its natural democratic order; this return to nature also brought about peoples' return to their senses—to each person's natural "common sense," which had been impaired by the artificial casuistry and irrationality, and brute force, of the monarchy.[27] In *Rights of Man* Paine draws America as not only close to nature but, as such, the place from which the world will be reformed:

> As America was the only spot in the political world where the principle of universal reformation could begin, so also was it the best in the natural world. An assemblage of circumstances conspired, not only to give birth, but to add gigantic maturity to its principles. The scene which that country presents to the eye of a spectator, has something in it which gen-

> erates and encourages great ideas. Nature appears to him in magnitude. The mighty objects he beholds, act upon his mind by enlarging it, and he partakes of the greatness he contemplates. Its first settlers were emigrants from different European nations, and of diversified professions of religion, retiring from the governmental persecutions of the old world, and meeting in the new, not as enemies, but as brothers. The wants which necessarily accompany the cultivation of a wilderness produced among them a state of society, which countries long harassed by the quarrels and intrigues of governments, had neglected to cherish. In such a situation man becomes what he ought. He sees his species, not with the inhuman idea of a natural enemy, but as kindred; and the example shows to the artificial world, that man must go back to Nature for information.[28]

So in what sense is democracy the "natural" form of political association? Here Paine, who believed in some kind of Superior Being but was suspicious of Christianity, nevertheless echoed Protestant conceptions of God, humanity, and the land, an echo of which he was, presumably, at least partially aware. It was an echo that also made his arguments resonate in an assertively Protestant America. Americans—indeed, people everywhere—were in direct, unmediated relationship with God and nature. In a statement evocative of the Covenant, Paine stated that "before any human institutions of government were known in the world, there existed . . . a compact between God and man, from the beginning of time: and that [man's relation with] his Maker cannot be changed by any human laws or human authority."[29] Elsewhere, he quotes from the biblical story of Gideon, who, turning away the Israelites' request that he be their king, says, "I will not rule over you . . . *The Lord shall rule over you.*"[30] Individuals were, in contact with nature, aware of their natural selves, their rights, and their relationship to God. Paine, like Protestants, rejected the idea of intermediaries with God—priests—or other intermediaries such as nobility. For Protestants God's word could be realized only in the manifestation of the ineffable—above all, in the Bible. For Paine, the Bible was a human document; the presence of a deity lay in nature alone. In either case, humans were in an individual relationship with their God or deities and as such in a state of equality with other humans. Government—legitimate government—in such cases could arise only through consent. (For the Quakers, Paine's religion of birth, even religious meetings were conducted in a way that recognized strict equality: they met in a circle so there would be no implied "head.")

This picture of democracy as the natural state of things was reinforced by Paine's use of the imagery of the people meeting in a democratic spirit under a tree.[31] And in his poem "Liberty Tree," addressed to American patriots in 1775, he wrote:

> In a chariot of light . . .
> The Goddess of Liberty came . . .
> She brought in her hand as a pledge of her love,
> And the plant she named *Liberty Tree.*
>
> The celestial exotic stuck deep in the ground,
> Like a native it flourished and bore;
> The fame of its fruit drew the nations around,
> To seek out this peacable shore.
> Unmindful of names and distinctions they came,
> For freemen like brothers agree;
> With one spirit endued, they one friendship pursued,
> And their temple was Liberty Tree.[32]

In this image of the tree as a place of congress, the thread that links the land, civic politics, and the people is powerfully invoked. The people are of nature and return to nature in the expression of their natural rights. These were divinely ordained rights, as nature itself was a creation and expression of God from the beginning of time, eliciting a certain Edenic sentiment.[33] The participants in this meeting under the tree are disparate individuals in Paine's moral tale in *Common Sense*—immigrants who had settled in some "sequestered part of the earth."[34] Implicit in Paine's story is that through deliberation ("under the branches") they become a people.[35] Democracy here is also of a local character, though Paine suggests that under the tree the "whole colony" could meet and deliberate.

What is striking in Paine's writings (like those of others of the time, such as Jefferson and Crèvecoeur, discussed below) is that the aesthetic beauty of nature—the awe and mysterious quality of it—and the rational and scientific predictability of nature were of a whole. It was the *orderly* character of nature that created a fusion of aesthetics and science, or an aesthetic beauty in science and rationality in aesthetics. Nature imbued with the sacred and nature as science were not in contradiction but, on the contrary, in synthesis. In "An Answer to a Friend," written in 1797 in a debate over Paine's the *Age of Reason* and seeking to refute arguments that the Bible was a divinely inspired document, Paine wrote:

> The Bible represents God to be a changeable, passionate, vindictive Being; making a world and then drowning it, afterwards repenting of what he had done, and promising not to do so again. Setting one nation to cut the throats of another and stopping the course of the sun till the butchery should be done. But the works of God in the Creation preach to us another doctrine. In that vast volume we see nothing to give us the idea of a changeable, passionate, vindictive God; everything we there behold impresses us with a contrary idea,—that of unchangeableness and of eternal order, harmony, and goodness. The sun and the seasons return at their appointed time, and everything in the Creation proclaims that God is unchangeable . . . The Bible represents God with all the passions of a mortal, and the Creation proclaims him with all the attributes of a God.[36]

Over the same issue in a letter to Thomas Erskine, a British jurist who defended Paine's *The Rights of Man* on charges of sedition, Paine exclaimed about "the Creator of the Universe, the Fountain of all Wisdom, the Origin of all Science, the Author of all Knowledge, the God of Order and of Harmony," and he then observed in a tone of awe, "We contemplate the vast economy of the creation, . . . the unerring regularity of the visible solar system, the perfection with which all its several parts revolve, and by corresponding assemblage form a whole . . . [in this] we trace the power of a Creator, [and his creations] from a mite to an elephant, from an atom to a universe."[37]

The rationality of nature allowed, by extension, for concepts of progress, technology, and economic growth. Natural rights, individual liberties, and economic "development" were all, so to speak, of the same cloth. Forms and patterns of nature were God's work, and to think scientifically and to cultivate and produce economically were, one can infer (in an echo of "Errand into the Wilderness"), to become a handmaiden of God, a maker of things in the image of the Maker. Thus we witness how the nascent nation-state was by its nature a "rational" project, whose mission came to be tied with science and technological and economic progress. Scientific rationalism and the sacred were conjoined. It was no surprise, in this context, that "men of science" were also "nation builders"—such as, notably, Benjamin Franklin. And Paine realized that the new America would need to be politically "invented," akin to technological forms of invention; this, necessarily, would be through legal forms.[38] That, in turn, would implicate America's understanding of its place in the world, geographically, politically, and morally.

The paradoxical combination of rationality with the sacred is critical to understanding American conceptions of "place" and its evolution. It is precisely, as Steven Grosby observes, "the formal rationalization of law which has historically been an essential factor in the formation of a 'people' and their 'territory.'"[39]

It is this combination, this dialectic, of scientific rationalism, with its stress on order, predictability, process, and progress and demarcation of the land, on the one hand and the sacred and primordial, even mystical, quality of the land on the other which plays out in Jefferson's writings. In his *Notes on the State of Virginia,* published in 1785, Jefferson begins by describing Virginia in exact, rational, terms delineating its precise latitudes and its geographic borders. The second paragraph is a description of the legal basis of these borders, noting that "these limits result from" charters, grants, treaty, and cession.[40] He also proffers descriptions of similar exacting fashion in answer to a series of queries on topography, economy, population, and so on. But this detached rationalist cannot but help celebrate what he sees (if not in such terms) as the primordial, sacred quality of the Virginian landscape and soil. On rivers, he describes the Ohio as the "most beautiful on earth" in the midst of describing other rivers in terms of depth and width, navigation and commercial uses. On mountains he talks of the Patowmac passage through the Blue Ridge as perhaps "the most stupendous scenes in nature . . . It is placid and delightful, as that is wild and tremendous." But Jefferson becomes positively entranced in describing the Natural Bridge. In a passage that begins by noting the bridge's dimensions, Jefferson writes that "it is impossible for the emotions, arising from the sublime, to be felt beyond what they are here: so beautiful an arch, so elevated, so light, and springing, as it were, up to heaven, the rapture of the Spectator is really indescribable!" In writing on slavery, where the slave has "to live and labor for another" and, as a consequence, their "morals" as well as industry are destroyed, Jefferson links God directly to nature, natural rights, and liberty. The proprietors of slaves become detached from nature, relying on others. In a much cited passage, Jefferson wrote, clearly in a shaken state:

> And can the liberties of a nation be thought secure when we have removed their only firm basis, a conviction in the minds of the people that these liberties are of the Gift of God? That they are not to be violated but with His wrath? Indeed, I tremble for my country when I reflect that God is just; that His justice cannot sleep for ever: that considering numbers, nature and natural means only, a revolution of the wheel of fortune, an exchange of situation, is among possible events: that it may become probable by supernatural interference! The Almighty has no attribute which can take side with us in such a contest. But it is impossible to be temperate and to pursue this subject through the various considerations of policy, of morals, of history natural and civil.[41]

It is a telling book, in its fusion not only of land, people, and politics but also of nature, science, and nationalism. It is also in large part in this book that he develops what became known as the agrarian myth. It was the feel, the touch, sight, and scent of nature that science revealed, not abstractions: "I am much indebted to you for [a] charming treatise on manures," Jefferson wrote in a letter to a friend, continuing, "Science never appears so beautiful as when applied to the uses of human life, nor any use of it so engaging as those of agriculture & domestic economy."[42]

This dialectic or interplay of the rational and the sacred is apparent in the writings of J. Hector St. John Crèvecoeur, a correspondent of Jefferson's. Crèvecoeur's *Letters from an American Farmer,* published in 1782, gained an extraordinary readership and following. However, in Crèvecoeur the dialectic is expressed in the link between the supposedly inherently democratic character of American agrarian life—"the richer soil . . . [that could] alter every thing; for our opinions, vices and virtues, are altogether local: we are machines fashioned by every circumstance around us"—and the rational organization of time. In Crèvecoeur, we find expression of the Protestant ethic, whereby the journey through space is a journey through time, progressing forward to freedom, prosperity, and civilization. Thus he draws on the archetypal story of the immigrant, from rags to riches, from oppression to liberty, from the poor soil of Europe to the verdant promise of America. The life of an American farmer is forward-looking and demands a methodical and sober outlook. Crèvecoeur declared in this vein, "After a foreigner from any part of Europe [has] arrived and become a citizen; let him devoutly listen to the voice of our great parent, which says to him, 'Welcome to my shores, distressed European; bless the hour in which thou didst see my verdant fields, my fair navigable rivers, my green mountains!—If thou wilt work, I have bread for thee; if thou wilt be honest, sober, industrious, I have greater rewards to confer on thee—ease and independence.'" And to an imagined Scottish immigrant he says, "Your future success will depend entirely on your own conduct . . . sober, . . . laborious, [and] honest."[43] In this we can see the appeal of the immigrant to American mythology; he or she is the American story, the story of the nation, writ small. The nation is on a journey in this historical saga, this rational, progressive movement through time, reflected in so many untold individual stories. The individual's journey represents a national epic.[44]

Furthermore, we see the rationalization of space, bounded, defined, and, as we shall see, carefully surveyed and mapped and intricately bound to the rationalization of time—methodical, progressive, forward-looking, carefully measured and calibrated, and predicated on political, economic, and social improvement. Put in simpler terms, this rationalization was expressed in the fundamental impetus to organize space and time according to a number.[45]

But there is a more significant point to be noticed in this combination or dialectic of the sacredness of place and its methodical, rational organization by borders, zoning, towns, boroughs, cities, rural areas, and so on. It is precisely by the process of geographic demarcation that the body politic is defined—borders are essential to designating a territory of some kind. In this apparently detached science of cartography and in defining the land in quantitative terms (not just borders, towns, and farms but rivers, mountains, and so on throughout the landscape), the earth is given its most palpable living form—a grounded manifestation of the infallible soul of the nation. A mountain does not know it is beautiful, a landscape does not know it is America (or France or Italy or Israel), a river does not know it is navigable; the topography is imbued with life through juridically defining it as, say, "Virginia." The land is defined and demarcated, and in this process "the people" are defined and constituted; indeed, they are given birth to by mother earth. Thus the critical role of law, often cast in highly technical, rationalistic, and arcane terms: processes of demarcation, jurisdiction, and "ins" and "outs" are all fundamentally legal processes. It is this dry character of law which hides its moral quality in constituting (literally—as in the Constitution!) the nation, its place, the individual, and his or her place in the nation, in the land. It is worth stressing here that this is a dialectical, or interactive, process, concerning the rational and scientific qualities, on the one hand, and the primordial and sacred qualities, on the other, of the land and its people; this is what Grosby apparently means when he argues that it is this "formal rationalization of law which historically has been an essential factor in the formation of a 'people' and their 'territory.'" He further observes:

> Indeed, the law of the land, which implies recognition of a bounded uniformity in subjection to the law—that is, what has sometimes been referred to as the formal rationalization of law—is a very important element in the establishment and, above all, the stability of the territorial boundaries of a society. This is the case because one element of the stability of a society is its consciousness of itself, that is, its self-image which guides its continual self-regulation and self-renewal. When a society is conscious of itself, the existence of that society is an object of the imagination and reflection of its individual members often through the contemplation of the events of its territorially bounded history, but also especially through the recognition of the legitimacy of the territorially bounded law of the society.
>
> . . . The connection between the rationalization of the law and the consolidation of the relatively extensive primordial structures of a "people" and a territory is just one of the many paradoxes of territoriality and its related phenomenon of the national state . . . Perhaps this paradox has become sharpest in . . . liberal democratic societies as they are constituted by both a recognition of the universal rights of man and a territorially delimited exercise of those rights.[46]

The extent to which law is so significant in delineating borders, shaping the landscape (through surveys, zoning, and the like), and, at the same time, determining the character of its members and their rights would be highly significant in the expansion of federal government and the slavery crisis. The expanding role of federal government in the very agglomeration of its law, ordinances, and legal presence would, it is suggested here, mold the federation of states that was the United States into a nation while also precipitating the sectional crisis. The dry "legalistic" patina of law obscures its essentially moral quality in defining authority, belonging, and, ultimately, right and wrong. (The present-day increase in the density of international law, notably international human rights law, needs to be understood in this light, and nineteenth-century America provides an interesting comparison. That is not to say that we are on a trajectory to "world union" but that the sociolegal and even moral landscape of state and nation is changing in ways that are referred to in Chap. 5.)

Imbuing the nation and the land with a sacred presence in space and time generated the rationalistic and scientific ethos, as, similarly, defining and delineating (in some quantitative sense) the "physical body," the land, of the nation-state was critical in "locating" it spatially, politically, morally, and historically. The nation was *designed* in two senses: it was sketched out cartographically, and it had a *design* in history, a "designated" role and mission. Initially, at least, it was viewed as a providentially inspired design—God drew up the charter.

The idealization of the pastoral scene was not new in art and literature, but the landscape's transformation as constituting the nation, and as part of a larger political design, was, states Leo Marx, by the late eighteenth century a newly articulated phenomenon.[47] Prior to this time "the pastoral" had been idealized for its uncultivated and aesthetic appearance. Thomas Jefferson played a critical role in the political recasting of the landscape in his articulation of the "agrarian myth," which informed American self-understanding and public policy, at least to the point of the "closing of the frontier" in the late nineteenth century. To a lesser extent, J. Hector St. John Crèvecoeur also played an important part in popularizing the idea of the independent farmer.

Landed Democracy

The Americans were conscious of themselves as a nation through sharing certain images—icons, pictures, symbols, literary ideas, founding documents—of what it meant to be, after all, an American. The building of monuments, parks, cemeteries dedicated to war dead, and the like would occur only with and following the Civil War (an issue we take up in the next chapter, and linked to a fundamental reevaluation of what "American" meant). The abiding symbol of antebellum America, in the North at least, is that of the independent farmer. The farmer represented the organic relationship with nature that was America. The farmer represented independence. The farmer represented sobriety, orderliness, hard work, methodical determination, and the Protestant ethic. The farmer was a thorough individualist. The farmer was civic minded, participating in and building his or her (occasionally the farmer was imagined as a woman) community. The farmer's life and community took place unmediated with those he or she knew face to face, in churches, schools, and businesses. The farmer lived in carefully organized (usually geometrically shaped) farms and organized time according to a rigid schedule as well. The farmer was forward-looking, "progressive." The farmer represented, above all, the land, and the land represented, for Americans, expansiveness, freedom, and liberty.

The farmer epitomized the "inextricable" link between land, civic self-government, and peoplehood, but one that reformulated Puritan notions. Where the Puritan writings reflect little concern with individuals as such but with the sacred relationship of the band of saints, of the New Israel, as a whole, with New England, the imagery from the Revolution is increasingly about individuals—the farmer, the frontiersman, the lone explorer (though those individuals symbolize stories the nation was telling about itself as a whole—including its sense of itself as a nation of individualists). Denominations, as noted earlier, increasingly reflected the idea that salvation was within the grasp of individuals (in contrast to Calvinist concepts of predestination), if only they chose a godly path, and religious organization also became much more decentralized and democratic. Thus, in contrast to Puritans maintaining a council of elders to ensure doctrinally correct actions by the body politic, for example, postrevolutionary America increasingly stressed individual rights. The Bill of Rights, religious freedom, and other rights progressively sanctified the role of the individual in American society—a phenomenon that became more marked as the nineteenth century wore on. This is related to the Second Great Awakening of the early nineteenth century, a religious revival that led to further erosion of Puritanism to more individualistic denominations such as the

Baptists and "New Light" Presbyterians and was also associated with abolitionist movements in the North and with women's rights.

The political tenor was, in other words, decidedly more liberal, and conceptions of "place" were mostly viewed as local. However, the increasing salience of individual rights would aid, as is argued further on in this chapter, the nationalization of government and identity. Related to this, of course, was the sectional conflict over slavery.

The focus on the farmer and local communities (after all, in this period, nine out of ten Americans lived in rural areas) was important in the early Republic in terms of what defined America as a union, but it also captured its contradictions.[48] Jefferson became the central icon of the agrarian myth and American republicanism and yet was claimed by both North and South in the ideological parrying over slavery.[49]

Nature for Jefferson, writes the political scientist Charles Miller, *was* America. Jefferson echoed and amplified the contrast with Europe—America's natural grandeur was set against European history, tradition, and monarchy. Like Paine and other figures of the Enlightenment, Jefferson used nature to attack traditional authority. Nature was an alternative form of authority—politically in the conception of natural rights—to that of kinship and feudal bonds. By extension, territory—rather than blood and kinship—came, in principle, to define community. Reinforcing the territorial aspect, American nature was not abstract, such as in the universal laws of physics, but was generally conceived of as about the palpable stuff of life—it was the nature of place, soil, plants, and animals. America was, in F. Scott Fitzgerald's evocative phrase, the "fresh green breast" of the New World. (The maternal references to nature reflect nationalism—the territory "gives birth" to the people in its constitution of them—and the etymology of the term, as well. The Roman goddess Natura "gave birth," as in "natal" or *natus.*) "Nature in the New World was a special truth that Americans intended to convert into a special virtue."[50] The image of the independent and republican farmer came to Jefferson, so to speak, naturally. Jefferson's sentiment about the relationship between the land and civic virtue—that is, a life devoted to the common good and a readiness in doing so to subsume one's private interests—is captured in a letter he wrote to James Madison, a coauthor of the *Federalist* papers and the fourth president of the United States, in 1787. American governments will, Jefferson wrote to his friend in a letter about the then yet-to-be ratified Constitution, "remain virtuous . . . as long as they are chiefly agricultural; and this will be as long as there shall be vacant lands in any part of America. When [the people] get piled upon one another in large cities, as in Europe, they will become as corrupt as in Europe."[51]

In a more famous passage from *Notes on the State of Virginia,* Jefferson compared agriculture with urban manufactures:

> The political economists of Europe have established it as a principle that every state should endeavor to manufacture for itself; and this principle, like many others, we transfer to America, without calculating the difference of circumstance which should often produce a difference of result. In Europe the lands are either cultivated, or locked up against the cultivator. Manufacture must therefore be resorted to of necessity, not of choice, to support the surplus of their people. But we have an immensity of land courting the industry of the husbandman. Is it best then that all our citizens should be employed in its improvement, or that one half should be called off from that to exercise manufactures and handicraft arts for the other? *Those who labor in the earth are the chosen people of God,* if he ever had a chosen people, whose breasts he has made his peculiar deposit for substantial and genuine virtue . . . Corruption of morals in the mass of cultivators is a phenomenon of which no age nor nation has furnished an example . . . Dependence begets subservience and venality, suffocates the germ of virtue, and prepares fit tools for the designs of ambition . . . While we have land to labour then, let us never wish to see our citizens occupied at a workbench, or twirling a distaff. Carpenters, masons, smiths, are wanting in husbandry; but, for the general operations of manufacture, let our workshops remain in Europe. It is better to carry provisions and materials to workmen there than bring them to the provisions and materials, and with them their manners and principles. The loss by the transportation of commodities across the Atlantic will be made up in happiness and permanence of government. The mobs of great cities add just so much to the support of pure government, as sores do to the strength of the human body. It is the manners and the spirit of a people which preserve a republic in vigour. A degeneracy in these is a canker which soon eats to the hearts of its laws and constitution.[52]

Jefferson was instrumental in teasing out the political implications of the symbol of the farmer. The cultivator of the soil was a palpable link between the land and the people, an expression of an almost Edenic notion that humankind was *of* the earth, and the quintessential narration, a story that could be conveyed so vividly in picture and in literature, of a *natural* right. It was the beginning and end of time, it was Genesis, where God created man from dust, creating an organic and interdependent relationship. It was in cultivating the soil, then, that humankind was whole, virtuous, and in its element, quite literally. "In the beginning, all the world was America," wrote John Locke in a book that deeply influenced Jefferson, as much as any other.[53] In this kind of imagination, this moral order, the cultivator of the soil, whether referred to as farmer, yeoman, or husbandman, became the pillar of republicanism and, more specifically, of the American Republic. The small landholders had both the economic and moral independence that was essential for republican self-rule.[54] In a letter to Madison, this one written in 1785, Jefferson recounted the pitiful state of a day laborer he met in France, and he attributed the oppressive condition of such laborers to the grossly skewered distribution of land in favor of the nobility. This was in his view a violation of a natural right, and it illustrated for Jefferson the importance of small landholders as the defining characteristic for a republican society:

> Whenever there is in any country, uncultivated lands and unemployed poor, it is clear that the laws of property have been so far extended as to violate natural right. The earth is given as a common stock for man to labour and live on. If, for the encouragement of industry we allow it to be appropriated, we must take care that other employment be furnished to those excluded from the appropriation. If we do not the fundamental right to labour the earth returns to the unemployed. It is too soon yet in our country to say that every man who cannot find employment but who can find uncultivated land, shall be at liberty to cultivate it, paying a moderate rent. But it is not too soon to provide by every possible means that as few as possible shall be without a little portion of land. The small landholders are the most precious part of a state.[55]

Similarly, in his *Notes on the State of Virginia,* Jefferson wrote that "cultivators of the earth are the most virtuous and independent citizens." (It is interesting how "cultivating" is metaphorically used more broadly to describe "productive" acts, as in Jefferson's desire that it "should be our endeavor to *cultivate* the peace and friendship of every nation.")[56] John Adams, the second president, argued that the "only possible way . . . of preserving the balance of power on the side of equal liberty and public virtue . . . [was to] make a division of the land into small quantities, so that the multitude may be possessed of landed estates." In manufacturing, Jefferson believed that workers would be dependent on the owners and whiplashed by the vicissitudes of the economy. Cities were locales of avarice and corruption, the place of stockjobbers, loungers, and other opportunists. If the independence of farmers actually promoted civic obligation, urban life had the opposite effect: a sense of selfish greed was endemic, a greed that translated into a lack of patriotism. "Merchants have no country," Jefferson wrote in a letter in 1814. "The mere spot they stand on does not constitute so strong an attachment as that from which they draw their gains." Lacking attachment to land and nature, they had no *place* in this world, an observation that perhaps could be made more so today, at least by a republican. Manufacturing produced, almost literally, rootlessness.[57] Jefferson's attack, in his *Notes,* on manufactures follows his attack on slavery, and it is clear that Jefferson saw slavery plantations as an obstacle to independent farms and hence republicanism (notwithstanding the widely observed irony of his slaveholding, derogatory comments on African Americans, and hidden relationship with a slave, Sally Hemings).

The link with nature is a recurring theme in Jefferson's thinking. From "Nature's God" that justified America's Declaration of Independence to the "chosen people of God," the cultivators of the earth, America the land, the people, the republic are indelibly sown together. The people were planted in the land and gained not only their sustenance from it but also their freedom.

The celebration of the land was not of a wild, primitive state but a rationalized, cultivated, controlled, and legally defined and bounded land. Thus the contrast was not simply between the corrupt city and the untainted wilderness. Indeed, there was much fear of the dissolute effects of the frontier, of subsistence farming, and of the savage state of the woodsman. Crèvecoeur talked of the "precious soil [that] feeds, clothes us [and upon] it is founded our *freedom,* our power as *citizens.*" He described how Americans "are a people of cultivators . . . united by the silken threads of government . . . all animated by the spirit of industry." Americans were like "plants, taken root and flourished!" He stressed the role of law in describing how oppressed peasants coming from Europe were transformed into active citizens and how, through the law, they were "naturalized"—the term *naturalized,* in this context, embodying both the legal and nature dimensions of becoming American. Americans, rooted in the land, loved their country, itself defined by the land. This was also a republican vision: this agrarian picture was one of civic obligations, civic restraint, and rational, sober, and orderly conduct.[58] Republican liberty rested in self-government, which in turn demanded civic virtue. George Mason, whose Virginia declaration of rights served as a model for part of the Declaration of Independence, observed in 1776 how civic virtue cannot exist without "frugality, probity, and strictness of morals."[59] (The planning of the West, as we shall see below, was designed to instill communal and public-spirited values.) Absolutist government was "wicked," but civic government, through active participation of citizens, was good. However, the complete lack of government, and of law and order, was feared. Nature, uncultivated and uncontrolled, could take on more ominous hues. Crèvecoeur suggested that the proximity of the woods created in solitary figures like hunters and frontiersmen—not civic-minded farmers—degenerate morals, viciousness, profligacy, and the absence of any communal spirit.[60] The Protestant ethic values represented by the farmer were viewed as critical for civic government and society; the work ethic was not yet associated with a corrosive conception of capitalism, or private enterprise, as selfish and greedy (except as represented in the cities and manufactures).

Jefferson, in expressing the civic virtues of an agrarian life and the dangers of urban industrial manufactures, was describing sentiments widely held in his America. (Such sentiments were nevertheless challenged by, most notably, Alexander Hamilton, and Jefferson himself softened his hostility later in life.)[61] Such sentiments held the view that every person had a right to land (a veritable "natural right"), that the cultivation of the earth brought title to it, that land brought not only independence but dignity, and that the cultivation of the soil also brought virtue and happiness.[62] Jefferson did play a pivotal role, in the context of this agrarian ideal, in opening the West to white settlement through the Ordinance of 1784 (discussed below) and, as president, facilitating the Louisiana Purchase in 1803, doubling the area of the United States, and exploratory missions, such as the Lewis and Clark Expedition, which surveyed the Louisiana Purchase and beyond to the Pacific from 1803 to 1806. Land thus symbolized a complex of factors: progress (in science and in expansion); liberty (in expansiveness); republican and civic virtues (in land as essential to independence and civic participation); and individualism (farmer, homesteader, and cowboy). It also served as a favorable contrast to the Old World (the pure, unsullied, and rural New World placed in relief against decrepit Europe).

Civic Localities

Man was of nature; his rights derived from nature. The pursuit of happiness could be attained only through his naturally given faculties. Thus man had a right to labor, which demanded possession of land (and contact with nature). It is the individual that is the center of this picture; it is the individual who is an organic relationship with the land. Nature, Jefferson determined, bequeathed the fundamentally equal worth of every individual. Jefferson's political vision derives from the individual and his rights (the male gender is used for substantive reason, noted below). From this conception of the individual as the anchor, the object, of civic society, government had to rest on the assent of those individuals. And given their equal worth, the only procedural mechanism for determining public sentiments, or policy, on different issues was majority rule.

However, what concerned Jefferson more than procedural issues were the substantive rights of the individuals concerned. Consequently he campaigned for a bill of rights guaranteeing the rights of individuals irrespective of majority will, and he expressed his displeasure to Madison in 1787 that the Constitution, as originally presented to the public, did not have a bill of rights.[63] In this concern for enumerated rights we see the basis of the enormous importance of the judiciary, and its historical growth, in American politics. Jefferson's wish for a bill of rights, as well as the wish of many others, would be fulfilled.[64] In fact, in state constitutions, more individual rights are enumerated, and more populations are included as rights-bearing, as the United States expanded westward—constitutions of the newer states had much more elaboration of rights than older states, expressing a linear, chronological process.[65] Over time, rights became more extensive not only in the United States but worldwide.[66]

Jefferson's focus on the individual was apparent in his discussion on immigration and citizenship. Just as civil government was based on the principle of consent, so must the admission of outsiders into citizenship be a matter of consent. Any society has the right to self-definition (what would later be referred to as national self-determination), which includes the right to determining its own citizenship laws. Any individual could exercise his or her right to expatriation, and Jefferson believed that slaves should be induced or forced to "exercise" that "right."[67] These conceptions of citizenship, not particular to Jefferson and indeed institutionally embedded in the modern state, reflect a largely unobserved phenomenon. Rights inhere *in place;* people who move in and out of a specific locale—say, the United States—implicate their rights. Territoriality is sacralized, even in avowedly liberal ideology such as Jefferson's. The individual is recognized as a social and political agent only insofar as he or she participates in a territorially defined nation. Or, to put it another way, citizenship is—or was, until recently—about belonging in a specific "space," and rights accruing to that citizenship are defined accordingly.[68]

Jefferson believed, reflecting and reinforcing a widely held belief, that such republican politics worked best at the "local" level. This view, it appears, arises from an "organic" conception of humans and nature, in which social relationships, community, obligations, and the like are expressed on a face-to-face level. Civic government similarly operates best and most virtuously when it is close to the people and is transparent in its actions. In fact, civic government should ideally be the civic embodiment of the people themselves, which is possible only if it is close to and "of" the people, in an almost literal sense. Social institutions that transcend the "natural" intercourse of individuals on a day-to-day level are thus, in some real sense, "artificial": Jefferson referred to these small units as wards, cantons, or hundreds. "Nature herself has confined, or limited," Jefferson copied down from a book he read, "the number of men in all societies that meet together to inform and be informed by argument and debate, within the natural powers of hearing and speech." Such communities could not exist in isolation, however, and when they joined with others to create a larger artificial association, federalist systems emerged. Jefferson was a believer in states' rights—the clarion call of the South during the slavery crisis—but for Jefferson states' rights derived from individual rights; southern apologists would later reverse that relationship.[69]

Jefferson's emphasis on natural rights had a peculiar twist, however, when it came to women. In the domain of the sexes "natural" took on an ascriptive or hereditary quality: thus Jefferson wrote, "Nature may by mental or physical disqualifications have marked infants and the weaker sex for the protection, rather than the direction, of government."[70] The contradiction in Jefferson's political approach is more stark here than on slavery; he saw the practice of slavery as theoretically and philosophically wrong, but women were "naturally" the "weaker sex," turning his own notion of natural rights on its head.

THE CELEBRATED French political philosopher Alexis de Tocqueville, writing in the mid-1830s, provided the contemporary understanding of American republicanism. There are, he wrote in arguing that states come before the federal government in the American body politic, "twenty four small sovereign nations . . . [and] the federal government . . . is the exception; the government of the states is the rule." The state, in turn, is made up of counties, and counties are made up of townships. It is in the township that, as Tocqueville presented it, true civic politics takes place. Much is made of Tocqueville's exposition of the role of voluntary associations in American democracy, but his discussion of townships—territorial units that served as the heart and soul of American republicanism—is less noticed. "The village or township is the only association," he wrote, "*that is so perfectly natural* that, wherever a number of men are collected, it seems to constitute itself," adding, "It is man who makes monarchies and establishes republics, but the township seems to come directly from the hand of God."[71] He used as his model townships in New England, which averaged two thousand to three thousand residents, a figure small enough, he said, that it ensured homogeneity of interests yet large enough that men with leadership talents could be found. These municipal institutions "constitute the strength of free nations." He went on,

> Town meetings are to liberty what primary schools are to science; they bring it within the people's reach, they teach men how to use and how to enjoy it. A nation may establish a free government, but without municipal institutions it cannot have the spirit of liberty . . .
>
> In the township, as well as everywhere else, the people are the source of power; but nowhere do they exercise their power more immediately [than in the township]. In America the people form a master who must be obeyed to the utmost limits of possibility.

As states cede powers to the federal government, so townships give up a proportion of their powers to the state; these units do not owe their existence to the federal government, as in France. (In fact, this idea that federal government was an act of states was shifting at the time Tocqueville wrote, a fact of no small consequence for the centralization of nationhood, as discussed below). The township is the node of civic life and society:

> The township, at the center of the ordinary relations of life, serves as a field for the desire of public esteem, the want of exciting interest, and the taste for authority and popularity; and the passions that commonly embroil society change their character when they find a vent so near the domestic hearth and the family circle.
>
> In the American townships power has been distributed with admirable skill, for the purpose of interesting the greatest possible number of persons in the common weal. Independently of the voters, who are from time to time called into action, the power is divided among innumerable functionaries and officers, who all, in their several spheres, represent the powerful community in whose name they act . . .
>
> . . . For in the United States it is believed, and with truth, that patriotism is a kind of devotion which is strengthened by ritual observance. In this manner the activity of the township is continually perceptible; it is daily manifested in the fulfillment of a duty or the exercise of a right; and a constant though gentle motion is thus kept up in society, which animates without disturbing it.

"The American," Tocqueville went on to say, "attaches himself to his little community for the same reason that the mountaineer clings to his hills, because the characteristic features of his country are there more distinctly marked; it has a more striking physiognomy." He then noted that "the native of New England is attached to his township because it is independent and free: his co-operation in its affairs ensures his attachment to its interests; the well-being it affords him secures his affection; and its welfare is the aim of his ambition and of his future exertions. He takes a part in every occurrence in the place; he practices the art of government in the small sphere within his reach; he accustoms himself to those forms without which liberty can only advance by revolutions; he imbibes their spirit; he acquires a taste for order, comprehends the balance of powers, and collects clear practical notions on the nature of his duties and the extent of his rights."[72]

Later on in volume 1 of *Democracy in America,* Tocqueville transposes the democratic merits of the township, by implication, to the state. He does this in the context of explaining the advantages of federalism. Large states tend to tyranny; republicanism tends to work best in small states. Federalism combines the two forms so as to promote political liberty while ensuring security from other large, predatory states. Here we observe how federal government is justified instrumentally. The federal union is not viewed, necessarily, as the "homeland," as the location of one's *place* in the world.

However, in the very instrumental approach to federal government a dynamic was set in place, largely unexpected, which generated the concept of shared nationhood and, in that process, contributed to the sectional crisis over slavery. We witness this in the Northwest Ordinance of 1787, which is predicated on Jeffersonian presumptions that civic republican societies were built on principles of equality and individual rights. Such principles are rooted in the land, literally inscribed through the surveying and design of townships built on geometric grids (representing equality and individual rights). But here is the irony: in using the federal government to develop public lands this way, and in using federal mechanisms to determine acceptance into statehood and the Union, the federal government became the locus for increasing accumulation of law—law that defined jurisdictions, principles, and rights—such that the federal level gradually came to represent the nation itself, to identify with or to rebel against. (The linkage between westward expansion, the expansion of federal law, and national identity is elaborated below.)

It is telling, in this context, that southerners objected to federal surveys and other federal measures even if on the face of it such measures did not invoke the institution of slavery. John Randolph, a prominent Virginian and member of the U.S. House of Representatives, in a speech in 1824 regarding a bill for surveys for roads and canals, railed that Congress by this bill would have enough power to "emancipate every slave in the United States."[73]

Furthermore, America's mission, what would come to be called its Manifest Destiny, was identified with the federal level of life—also contributing to the secession of the Confederacy. Thus, insofar as republican ideals of equality and individual rights were emphasized, it was on the constitutional and federal level that this mission was institutionalized, as is especially notable in the Northwest Ordinance of 1787. Many, including Abraham Lincoln, could accept that the America of the present moment had its flaws, above all slavery. But those same people were imbued with the notion of America's mission to reform itself, and indeed the world, over time. "A sense of mission" to redeem the world through the promotion of government by consent, republican in character, free of hereditary rule, characterized much of the country, wrote the historian Frederick Merk; it was a thread that ran from the Puritans through the nineteenth century and, indeed, much of the twentieth century as well.[74] The sense of history was "progressive," even if there was a readiness to make compromises with the present. Thus it was the *expansion* of the United States that rubbed the sores of discontent between North and South.

The Northwest Ordinance of 1787

The Northwest Ordinance graphically illustrates (literally) the way certain ideas of republicanism and nationhood were sown into the land and how the American sense of place came to be nationalized. The ordinance tends to be marginal in the historical consciousness, however, and thus needs to be briefly described before turning to its inscriptive role.

Having defeated the British in the War of Independence by 1783, the thirteen states, represented in the Continental Congress, were confronted by the issue of the Northwest Territory. That territory constituted the frontier zone lying west of Pennsylvania, east of the Mississippi River, north of the Ohio River, and south of the Great Lakes. Several practical issues concerned the Continental Congress with regard to the Northwest Territory: The territory posed security questions. Frontiersmen, viewed as unruly and untrustworthy, could provoke war with Native American tribes. Similarly, there was danger of war with the British, Spanish, or even the French, who were present in the border regions, in a struggle over land. The Congress also feared that the power vacuum in the Northwest, following the expulsion of the British, would lead to competition between the loosely confederated states, essentially sovereign entities prior to the ratification of the Constitution in 1788. Furthermore, the United States was in desperate need of money, having incurred debts in the independence struggle, and new lands could be a source of revenue.[75] Until roughly 1780 the Northwest Territory lands had been claimed by existing American states, after which they ceded the lands to the United States government.

The Continental Congress enacted three ordinances, in 1784, 1785, and in 1787, to deal with the Northwest. The Ordinance of 1784, which Thomas Jefferson had drafted, created a number of self-governing districts in the territory, each with their own representative in Congress upon reaching a population of twenty thousand. Each district would become a state once its population was at the level of the least populous existing state. The 1784 ordinance was superseded by the Ordinance of 1787. The Ordinance of 1785 required the systematic surveying of the territory. That survey divided the land into a rectangular grid system. The basic unit was a township of six square miles. The township was then subdivided into rectangular plots for individual ownership. The minimum parcel of land was one square mile, or 640 acres, at a minimum cost of one dollar per acre. Each township was required to set aside land for a school. This grid system guided land policy until the Homestead Act of 1862.[76]

The most important of the three ordinances was the Northwest Ordinance of 1787, which described the form of government of the districts (the constitutive parts of the Northwest Territory) and the basis for statehood in the Union. Congress appointed a governor and judges for each district until it reached a population of five thousand free adult males, after which it became a "territory" and formed its own legislature. A population of sixty thousand was necessary for statehood and admission into the Union. The ordinance also outlawed slavery in the Northwest Territory; guaranteed freedom of religion and other civil liberties; promised fair treatment of resident Native Americans; and provided for education. These districts were promised, once passing the population threshold of sixty thousand, equal status with existing states, precluding the ongoing subservient status of a colony. The Northwest Territory would, it was stipulated, eventually comprise at least three but not more than five states.[77]

Few Americans, let alone foreigners, are familiar with the Northwest Ordinance of 1787. But its historical and sociological importance, it can be argued, is nearly equal to that of the Constitution and in respects (such as the slavery issue) goes beyond the Constitution. Legally, some have argued for it having the same quasi-constitutional status as the *Federalist* papers or the Declaration of Independence. It set the framework for territorial government and statehood which applies to thirty-one of the fifty states in the Union (it served as a model for states beyond the Northwest as defined in 1787). The ordinance provided a blueprint and was the foundational document for expansion, as well as a political model for governance on the state (and, by extension, federal) level. It defined the process whereby most states would join the Union, thus playing a critical role in shaping the nation. It helped preserve the newly won national political identity by addressing threats, or perceived threats, to its security. The ordinance, as is discussed further below, would prove critical in maintaining and strengthening the Union and in challenging centrifugal forces. It sought to define jurisdictional areas clearly, so as to avoid disputes over title. It helped sell western lands, ensured sanctity of contracts, and provided order in area widely seen as dangerous and chaotic. It also encouraged the settlement of sober, civic-minded, and entrepreneurial citizens, which was understood as critical for the ultimate establishment of republican states.

Perhaps most strikingly, however, the ordinance articulated principles that are not explicated in the Constitution. In a curious way, the Constitution is a more procedural document in its description of government (reflecting, presumably, its more instrumental role referred to earlier): the ordinance is more descriptive of the raison d'être and the social vision of America than the Constitution, prior to its amendments, was. The ordinance banned slavery and involuntary servitude (and was celebrated as such by abolitionists); encouraged expansion in a Union framework, promoted civic education, endorsed the centrality of private contracts, sought good relations with and education of American Indians, and directly promoted "public virtue." On issues such as freedom of religion, the ordinance was much more moralistic in tone than the First Amendment of the Constitution, stating in Article I, "No person, demeaning himself in a peaceable and orderly manner, shall ever be molested on account of his mode of worship or religious sentiments." Similarly, in Article III it states (revealing the clearly republican and civic sentiments of its authors) that "religion, morality, and knowledge, being necessary to good government and the happiness of mankind, schools and means of education shall forever be encouraged." States that were formed out of the Northwest Territory, it was noted in an almost prescient expectation of the sectional crisis, "shall forever remain a part of this Confederacy of the United States of America."[78]

The Northwest Ordinances are of special interest here because it is—to an extent we do not see in the Constitution or the Declaration of Independence—in these ordinances that the particular forms of association between people and place were debated, defined, and framed. The agrarian myth and nascent American ideas of place were *inscribed* into the land in a vivid fashion through surveying, mapping, zoning, making townships the organizing principle of spatial organization, stipulating the shape and size of farms, and so on.[79] Myths were imprinted on the land so as to become "material" and to be realized in social and political practices. Thus, there was no "objective" moral association between people and place, as such, but a certain association with place was realized (as in "made real") through *cultivating* the land in terms of socially, economically, and politically prescribed ways. In etching out a certain moral geography, so lines are drawn constraining, shaping, and channeling future developments, and such "borders" become markers of conflict as well. This process also highlights the centrality of law. The ordinances themselves were, of course, legal acts and documents, but in a larger sense law—the legal process—becomes the medium whereby morals and ideas are formalized, contested, and negotiated. Through lawmaking people define their relationship to the land and to one another.

What the Ordinance of 1787 (in tandem with the Ordinance of 1785) graphically inscribes is the centrality of geography, of a bounded territory, in constituting community. American political and social life was arranged not only geographically but locally, with the township, the "natural" extent of a community, serving as the hub of spatial organization. The ordinances also follow the dictum of Jefferson (most notably, but also others, such as John Adams) that republicanism and public virtue were best served by dividing the land into relatively small and equal parcels so the "multitude" could possess their own "landed estates." This land would, it was understood in the Jeffersonian and republican sense, provide independence not only for sustenance but for freedom. Thus, the imagery of the independent farmer shone through the ordinances. The land was celebrated but also "rationalized" through careful surveys, zoning, and geometric organization (the identical grids also representing equality). The land was bounded and ordered.

The requirement that in each township land be set aside for public schools derived from the republican desire to inculcate civic virtue in the residents, a proactive concept of citizenship notably absent in the Constitution. Indeed, the Ordinance of 1787 pictures "the" United States that need (the plural being used here advisedly) to be bound together through common interests, not through common nationhood. The political community was still conceived of as lying at the state level. Thus, though incipient expressions of nationhood at the federal level are heard even before the ratification of the Constitution, one's land was still viewed largely as that of one's state. Jefferson's vision of republicanism as working best in small units was by and large written into the ordinances. Indeed, the Ordinance of 1784 made the future states of the Northwest so small that it was feared they would be overshadowed by the older, established states. Under the 1787 ordinance, their size was increased. Still, political community was seen as "local," not national.

Legally, the location of political community is revealed in the understanding of citizenship in the 1787 ordinance. Under that ordinance, residents of a territory attained full citizenship rights only when the territory became a state. Under district and territorial government, the residents only had a partial citizenship, with a governor initially appointed by Congress and without any representation in federal government even with the "territorial" status. This highlighted the importance of states: full citizenship was derived from membership in a state; it was not an ipso facto status due to residency in the territorial domain of the United States. Tellingly, the courts in antebellum America could not or would not determine whether state citizenship or citizenship of the Union was preeminent.

Another understanding that came out of the Ordinance of 1787, namely, that it was a "compact" between the United States and the settlers, reinforced that states were the location of political community. The idea of the compact was to indicate that the territories were not colonies with a subservient status vis-à-vis the nation but rather discrete and even autonomous communities (represented by the respective states) in a federated but, in principle, equal relationship with one another. Nationalists attacked the compact notion, arguing that the United States could not make a compact with "themselves." Many believed, however, that it threatened republicanism to extend the "political community" beyond the level of the state.[80]

And yet the federal government had a critical role in this republican system, not only in terms of the division of power between states and the federal government but also in guaranteeing the rights of the individual. The Constitution reinforced what would prove to be a pivotal role of the federal government in its Article 4, Section 4, guaranteeing every state in the Union a "Republican Form of Government," though in the Constitution the defense of individual rights would be limited to whites in its first eighty or so years. The Ordinance of 1787 went beyond the Constitution, which it marginally predated, in its ban on slavery in the Northwest.[81]

It is also in the ordinances that one can observe the almost seamless weaving of the science of surveying and mapping with the moral fabric of American republicanism. The surveying determined the boundaries of the body politic, its "geographical body." The townships, in effect, became the cells of a larger entity as they were defined and enclosed by that larger, federal and national landscape. It was also felt that "rationalizing" and demarcating the landscape in this manner would foster good, civic-minded, upstanding, and sober stakeholder-citizens. The stress on education for the promotion of civic values would similarly serve this purpose. Citizens' private interests would thus be harnessed to serve the public good; this was in contrast with the uncivilized, uncontained, violent, non-civic-minded woodsmen and other such riffraff who lived in a virtual state of anarchy. The rationalization of the landscape created reasoned and rational citizens, in contrast to the wild and untamed woods and its wild and untamed progeny.[82]

Locating the Nation's Frontiers

Frederick Jackson Turner presented his frontier thesis at a meeting of historians in 1893, suggesting that the frontier and expansion westward had developed the distinct American character. He developed this thesis by first noting a bulletin of the superintendent of the census for 1890 announcing, in effect, the closing of the frontier. Turner's essay on the frontier is repeatedly referred to as the most singularly influential thesis and interpretation of the American past. What has stuck most in Turner's thesis is what he himself termed the "most important effect of the frontier." That effect had been the promotion of democracy and the promotion of American individualism, due to the primitive conditions of the wilderness.[83] Though his thesis lives on, the linkage with democratic individualism has been criticized—for one thing, frontier conditions elsewhere in the world, such as South Africa or Canada, did not produce the same democratic outlook or "rugged individualism."

Two aspects of the thesis are, however, of interest here. The first concerns the idea that geography molded the American identity: that this idea had such a resonance in the public is indicative of how geography plays in Americans' imagination of themselves.[84] The great open spaces, the sense of nature itself, conjure up a nostalgic sense of freedom. It is expressed in American literature and also in American music, as in Van Morrison's song "Brand New Day": "When all the dark clouds roll away / And the sun begins to shine / I see my freedom from across the way / And it comes right in on time." The other aspect is a subtle argument that Turner makes about the forming of a national identity which has been largely overlooked, namely, the linkage between expansion westward and the expansion of federal statutes and law. In Turner's argument, at least by implication, the growth of the federal presence through law molded, in turn, American nationhood: "The legislation which most developed the powers of the national government, and played the largest part in its activity, was conditioned on the frontier. Writers have discussed the subjects of tariff, land, and internal improvement, as subsidiary to the slavery question. But when American history comes to be rightly viewed it will be seen that the slavery question is an incident . . . The growth of nationalism and the evolution of American political institutions were dependent on the advance of the frontier."[85]

In saying that the slavery issue was an "incident" Turner overstates his case; but slavery did indeed become more "marked" as the federal legal, topographical, jurisdictional, and moral grid expanded (as in the Northwest Ordinance of 1787)—and thus the sectional conflict became especially acute when issues of territorial expansion arose. (This is also revealed in the southern opposition to federal surveys and internal improvements noted earlier.) Turner recognizes this when he referred to the lands in the public domain:

> The public domain has been a force of profound importance in the nationalization and development of the government. The effects of the struggle of the landed and the landless States, and of the Ordinance of 1787, need no discussion. Administratively the frontier called out some of the highest and most vitalizing activities of the general government. The purchase of Louisiana [in 1803] was perhaps the constitutional turning point in the history of the Republic, inasmuch as it afforded both a new

> area for national legislation and the occasion of the downfall of the policy of strict construction. But the purchase of Louisiana was called out by frontier needs and demands. As frontier States accrued to the Union the national power grew.

Turner then notes, quoting from a dedication speech for the Calhoun monument: "In 1789 the States were the creators of the Federal Government; in 1861 the Federal Government was the creator of a large majority of the States." He adds, "When we consider the public domain from the point of view of the sale and disposal of the public lands we are again brought face to face with the frontier . . . It is safe to say that the legislation with regard to land, tariffs, and internal improvements . . . was conditioned on frontier ideas and needs."[86] Thus, the territorial expansion of the United States expanded the presence of the federal government—through legislation and, as a consequence, its jurisdictional role—and this was critical in the nationalization of American identity.

An analysis of the *Statutes at Large,* the congressional record of public and private laws, is one gauge of the growing legal presence of the federal government, particularly in the area of federal public land laws. In the first decade of its existence, through 1799, Congress enacted 14 public land laws; 48 public land laws from 1800 through 1809; 71 such laws from 1810 to 1819; 117 enactments in the decade to 1829; 107 public land laws in the decade to 1839; 110 from 1840 to 1849; and 188 public land laws from 1850 to 1859, on the eve of the Civil War. Perhaps even more significantly, public land laws are a growing proportion of all federal public acts in this period. In the first decade after 1789, they are less than 3 percent of all public laws, but they grow steadily as a proportion of public laws in almost every decade up to the Civil War. From 1840 to 1849, public land laws constitute more than 18 percent of all public laws, and in the decade after 1850 public land laws are almost a third, some 32 percent, of all public laws passed by Congress.[87]

The slavery crisis was exacerbated by this growing presence of the federal government, a situation that was repeatedly evident in various conflicts over what to do with new territories and in determining whether new states would be free states or slave states. The expanding federal presence includes, as noted earlier, not only its growing legal presence in the narrow sense but also its critical role in the understanding of America's mission or Manifest Destiny.

This identification of federal government with a forward-looking, redemptive role is revealed in Lincoln's objections to compromising on the slavery issue *outside* the recognized slave states, as territorial expansion concerned the United States' preordained destiny to expand liberty. In 1860 Lincoln was president-elect of a party that had grown out of disputes over extending slavery, and it was a party that was strictly northern in its sentiments. To save the Union as the rumblings of war grew louder, compromise proposals were put forward, including one that would create a constitutional amendment to allow slavery in certain territories. Lincoln was adamantly opposed. On January 11, 1861, he wrote: "Either way, if we surrender, it is the end of us and the government." The southerners "will repeat the experiment upon us *ad libitum.* A year will not pass, till we shall have to take Cuba, as a condition upon they will stay in the Union." On February 1, he wrote to William H. Seward, who would become Lincoln's secretary of state a year later: "I say now, however, as I have said, that on the territorial question—that is the question of extending slavery under national auspices—I am inflexible. I am for no compromise which *assists* or *permits* the extension of the institution on soil owned by the nation. And any trick by which the nation is to acquire territory, and then allow some local authority to spread slavery over it, is as obnoxious as any other."[88] Lincoln was against *extending* the institution of slavery or allowing *local* authorities discretion on that matter. It was the federal government that represented America's Manifest Destiny, its path into the future.

Landscape Architecture

The anthropologist James Scott writes in *Seeing like a State* that he began his study by asking why the modern state is the enemy of "people who move around." The more he examined efforts at making populations sedentary, the more he came to see them as attempts by the state to make the population "legible." Premodern states were partially blind: they knew little about their subjects, their landholdings and wealth, their location, or even their identity. So how did the state get a "handle" on its subjects and environment? Here Scott sees a disparate set of measures designed to make the population and landscape more simple and legible, measures that would quantify, standardize, "make measurable," and, overall, rationalize the environment. These measures included surveys (like the ordinances of 1785 and 1787) and maps, population registers, and generally administrative ordering. Once the state had legibility and simplification, it had control and the ability to manipulate the society over which it ruled. (Or thought it had; Scott notes that the most tragic episodes in twentieth-century history arose out of attempts at massive social engineering, such as Nazi Germany, Maoist China, or Stalinist Soviet Union.)[89]

To control the people and property the state made use of, respectively, citizenship and a legible property system. In the latter case, legally deeded, measurable, and uniform forms of property registration were necessary, as opposed to the premodern hodgepodge of local customs, shared lands, vague boundaries, and disparate forms of measurement. Control over the population was promoted through homogeneous citizenship. In place of different estates and different status groups (monks, aristocrats, peasants, and so on) unequal in law, the singular and equal citizen would be the bedrock of a national society. Citizens could thus be added up, measured (such as in the census), controlled (as in passports), mobilized (for the military), and categorized (for schooling purposes, for example). The now imputedly monolithic individual became a legible demographic category and as such relatively easy to manipulate. In place of a myriad of incommensurable small communities, a national society would arise, where any citizen could fit in anywhere.[90]

Scott's argument is a compelling one, but he overstates the extent to which purposive "state control" is universally at the center of this process and certainly overstates it as far as the American case is concerned (he notes the surveying and mapping under the ordinances as an example of his argument). As we have seen, the idea behind the Northwest Ordinance was to institute republican control, and that was understood to take place at the local level. In other words, Scott looks at the process of rationalization—quantification, measurement, standardization, and simplification of the population and the land—as an overwhelmingly "top-down" process driven by state desire for control. He largely overlooks the cases in which this process of rationalization was "bottom-up"—when this process of rationalizing the population and landscape was a means to ensuring individual and civic rights vis-à-vis the state.

However, the irony is that the *outcome* of instituting local republican control was to nationalize American society, ultimately to make an American nation. Through this Jeffersonian mapping and "landscaping"—physically, socially, politically, and morally—republican *and* individual prerogatives came to rest in the federal government and vice versa; a dialectic, a relationship, was set in place which, if unintentionally, generated a national frame of defining where, ultimately, citizenship loyalties lay and what constituted "an American." In this context, the conflict over slavery was indeed unavoidable. In that regard, Lincoln was right: no compromise with the South over slavery could have endured.

A Country's Reservations

Surely a painful irony of American history is this: the very way the land was inscribed, through the agrarian myth, and its starkly different moral geography, which would lead to the clash with the South over slavery, *would also lead to the destruction of precolonial American Indian culture.* Native Americans, it was understood, should be converted to farming, to live sedentary rather than nomadic or pastoral lives, and to live in bounded, mutually defined exclusive plots of land and reservations. "We want to make citizens out of them," wrote one government architect of American Indian policy in a revealing statement, "and [in order to do so] they must first be anchored to the soil." American Indians, in moving to a "higher stage" of civilization, would be educated and civilized.[91] This same moral geography could later be celebrated in the confrontation with slavery while bemoaned in its treatment of native peoples; but that moral geography had its own internal logic, which would destroy other forms, regardless of what, in retrospect, we see as their moral worth.

Nature's Nation: Preserving the Future

THREE

If we are mark'd to die, we are now
To do our country loss . . .
From this day to the ending of the world,
But we in it shall be remembered—
We few, we happy few, we band of brothers;
For he today that sheds his blood with me
Shall be my brother.

—WILLIAM SHAKESPEARE, *Henry V*

ON NOVEMBER 19, 1863, Abraham Lincoln, speaking to an audience of some fifteen thousand, made the dedicatory remarks that became known as the Gettysburg Address. The soil and the bodies of lost Union soldiers at Gettysburg intermingled, symbolizing civic duty and sacrifice and the roots of a reborn people—thus Lincoln took that bitter sacrifice and spoke it into a national story, a poem, of faith and redemption. The blood of the martyrs fed the soil of the nation: the land, the people, and republicanism once again became one, but the Union was now an unbroken and singular landscape. The Declaration of Independence had created these United States. The Gettysburg Address, delivered during what some had termed the Second American Revolution, marked *this* United States. What Thomas Jefferson was to the early Republic, Lincoln was to the Republic's "*new* birth of freedom," each in their respective ways turning to nature, to the land, to constitute and shape the people; as in the metaphor of John Cotton in his sermon before the Great Migration of 1630, the people were—had to be—*planted* in the land. The Gettysburg Address became a monument, a moment made eternal. The address was etched into the nation's understanding of itself, as much as the battlefield and the cemetery were imprinted into the ground. Indeed, the literary and physical retelling and recasting of that landscape as a national park, monument, and place of pilgrimage were critical to securing a common and shared America.

The sectional conflict had demonstrated painfully that Jeffersonian agrarian democracy had failed—failed in that Jefferson's idea that republicanism and "localism" were intertwined was patently wrong. The proslavery forces had tried to shield the institution under the banner of states' rights, pointing to centralized government as a threat to liberty. Indeed, Jefferson's writings were selectively used to buttress this point. (Abolitionists also selectively drew on Jefferson, more in terms of his support for natural rights and equality.)[1] No one was, even after the Civil War, about to disavow Jefferson. Nevertheless, his belief in republicanism and equality could not be maintained with his belief that civic politics worked best at the level of relatively small communities—at least not in the South. And his agrarian idyll was quickly coming up against a country that was rapidly industrializing and urbanizing in the final third of the nineteenth century.

The contradictions of the Jeffersonian vision were apparent early on in the sectional conflict. In a debate on congressional requisitions for western surveying in 1830, when proslavery and antislavery factions contended for dominance, Daniel Webster replied to a spokesman of the South in a packed chamber of the Senate, in a speech that Lincoln considered the greatest American oration ever. Webster, who had a long career as orator, lawyer, politician, and secretary of state as well as senator, concluded the speech with a line so compelling, so at the crux of the conflict, that it has resonated ever since: "Liberty *and* Union, now and forever, one and inseparable." (Although the issue of surveying, which was the backdrop to the debate, has been termed by one historian as "trivial" in the context of the slavery debate, in fact in a certain sense surveying gets to the heart of the matter—this struggle was about mapping the landscape, and that mapping was inherently moral, as well as physical.)[2]

A republic of equals could not survive without Union. But the Union was also of the land, of the soil, and at Gettysburg that soil would nourish a new nation. The nation still had to be planted in the land; yet nationhood could not depend on the image of the independent yeoman farmer, in palpable touch with that land. Nor could community, in any national sense, be face to face or local. The link to the soil, the very body of the body politic in a republic of equals, the organic core of the nation, and the link to a national community of millions had to be *mediated.*[3] The farmer's connection to the land was "immediate" (meaning without mediation); post–Civil War America had to be built differently. Yet the link with the land remained critical—republicanism could not be otherwise. Indeed, "natural laws" and nature became even more accentuated in the crushing of slavery (which, in its kinship-based, castelike quality, Lincoln associated with monarchy).[4] Thus we see, beginning with Gettysburg, the rise of national monuments and parks—they were symbols, in the sense of "mediators" or media,of the link to the land and, in marking public spaces, defined how the nation imagined itself, geographically and historically.

Such monuments and parks were the objects or structures through which the nation could imagine itself a nation, could be "conscious" of and reflect upon itself as a nation. Monuments delineated the space, as well as the tie to the land, through which citizens individually would understand the bounds of (their) nationhood. They captured the "spirit" of the nation in the way art captures emotion and in this regard expressed the dialectic of nature as material artifact and nature as constituting life itself. War monuments and cemeteries in particular, through commemorating civic sacrifice, gave a palpable sense of the link between the body of the nation (through the bodies of lost soldiers) and the soil. The widespread sense, as the nineteenth century came to an end, of the "close of the frontier"—and what this implied for the vision of a democracy of farmers and endless open land—reinforced the mediative role of national parks for nationhood.

The monuments and parks presented a new story of America's role in history, but one that sets up the frame for struggles over national self-definition that would come much later. The monuments and the parks become the touchstone (almost literally) for defining the United States.

Sweet Land of Liberty: Sacrifice, Soil, and Freedom

Lincoln called up a sense of the land, evoking it as the source of life of a nation "conceived" and "brought forth" in a "new birth of freedom," and of the earth binding together a "naturally" singular nation.[5] This sentiment about the land was not foreign to his audience at Gettysburg. The speech resonated with his listeners and with those who subsequently read the address in newspapers, as it articulated the place of nature in a way that was palpable to many nineteenth-century Americans. The cadence of life and death, and the liminal quality of the cemetery, surrounded by the green of nature, were likewise themes that echoed in the minds of his audience.[6] Nature and rural "garden cemeteries" were sites of Protestant-like notions of *unio mystica,* where God could be experienced, where heaven touched earth. (The "experience" here was understood to be more emotional or mystical than intellectual; in the nineteenth century many sects and denominations had broken away, to varying degrees, from the cool rationality of the Puritans.)

The visual grandeur of the land and nature, and the national vista it conveyed, were expressed in landscape painting, a major totem of collective identity from 1825 through the mid-1870s. The period of landscape painting's relative decline may be significant—it is roughly at that point that national parks and monuments began to be established as the markers of nationhood.[7]

Those who sought to create a national art were consciously seeking to ground—literally—a people who were not yet a nation but, in the words of a traveler in 1839, "a mass of people cemented together to a certain degree, by a general form of government." As the art historian Angela Miller writes, nationhood and national identity could not be built on abstract political principles alone but needed an *organic* basis, and the environment, the living soil, became that basis. This was an ongoing theme in American history, but landscape art now had the express purpose of presenting the landscape as a harmonious whole, which suggested by extension a singular people. It also reflected that a republic of citizens, in the absence of kinship ties, had to be rooted in a bounded land. (Interestingly, when the term *organic* is associated with nationalism, it is usually paired with *Volksgemeinschaft,* that is, nationalism based on blood descent together with territoriality, as has historically been the case with German national identity.) The landscape artists were informed by northern sensibilities; they found the South repugnant aesthetically as well as politically (and rarely ventured there to paint). Thus their paintings reflected northern "ways of life" but were designed to portray the whole nation writ on a single landscape painting. The artists, such as Thomas Cole and Fredric Church, represented a northeastern, but nationally minded, elite.[8]

A unified nation had to be first imagined, since the natural world in its infinite complexity defies empirical description (our words are, after all, massive simplifications), and these Hudson River landscape artists, as they were known, gave powerful visual testimony to the American drama. Their paintings incorporated themes in such a way that the landscape was not only presented as spatial but also told of America's temporal and historical calling, its Manifest Destiny. Farmers, frontiersmen, and other Americans and the advance of the frontier and civilization were all depicted as part of the natural environment. Later, in the era of national parks, the ideal of nature was depicted in photography and literature as pristine, pure, and free of human "intrusion" and even "contamination."

In the Hudson River School the landscapes were also frequently anthropomorphized, with natural phenomena (clouds, rocks, and so on) given a human or divine countenance.[9] The portrayed scenery also evoked in many cases maternal or sexual imagery. The nation and nature have customarily been referred to in the feminine gender, as if to convey "her" as mother earth, the place of birth of the nation, and the nation itself as the motherland. As a minister at the time expressed it, "Grand natural scenery tends to permanently affect the character of those *cradled in its bosom,* [and] is the *nursery of patriotism* the most firm and eloquence the most thrilling." One writer in the 1840s, drawing on sexually resonant imagery, described the effect of "atmospheric landscape" paintings in the following terms: "The settling of the quivering balance, the ultimate swell of the choir, the mellowness of full moontide, the entire calm that succeeds both excitement and reaction—in a word, that completeness, satisfaction, content, which like the calm glow of autumn, seems to fill all conscious desire, and hush the pleadings of expectancy . . . an equilibrium of the soul."[10] A critic, writing in 1866 of Sanford R. Giffords' *Hunter Mountain, Twilight,* described it as a "nude"—a painting of a mountain that resembled a nude woman lying on her back. This was a metaphor of mother nature, the art historian Gray Sweeney notes, though after the Civil War man's presence was viewed more somberly as a scar on the landscape.[11] F. Scott Fitzgerald would later draw on such imagery of nature in his novel *The Great Gatsby,* memorably invoking a scene in which the first European settlers first see the "fresh, green breast" of the New World from their ships. The anthropomorphized depiction of the landscape suggested a vital quality and spiritual presence, the symbiotic relationship between nature and humankind, and the handiwork of the Creator.

The paintings of the landscapes, of nature, invoked nationhood and the divine more forcefully than nature, in its actual form, could. The art historian Barbara Novak describes the palpable linkage of nature, God, and national community, of the "community through nature," evoked by these paintings: "God in or revealed through nature is accessible to every man, and every man can thus 'commune' (as the word was) with nature and partake in the divine. God in nature speaks to God in man . . . And man can also commune with *man* through nature—a communing which requires for its representation not the solitary figure, but two figures in a landscape . . . The sense of community fostered by the natural church was reinforced by an all-pervasive nationalism that identified with the American landscape." Novak writes further that "the unity of nature bespoke the unity of God . . . Thus the landscape painters, the leaders of the national flock, could remind the nation of divine benevolence and of a chosen destiny by keeping their eyes before the mountains, trees, forests, and lakes which revealed the Word in each shining image."[12] That unity of nature bespoke not only the unity of God but also the unity of the nation—nature was the soul and body of the nation, and the promise of history and redemption. To break that unity (as the dark clouds of the sectional crisis threatened) was to desecrate Providence and national destiny.

The human and anthropomorphic depiction in these landscapes hinted at the divine, the maternal, and also romantic love. Thomas Cole, writing of trees, noted that "they spring from some resemblance to human form. There is an expression of affection in intertwining branches . . . [Trees] assimilate with each other in form and character." Elsewhere he wrote that the "dark cavern is emblematic of our earthly origin and the mysterious past . . . [whereas the] rosy light of morning, the luxuriant flowers are the emblems of the joyousness of life."[13]

Angela Miller suggests that after midcentury landscape painting was not "exclusively defined by visual forms that translated expansionist ideology into the structured experience of a humanely ordered nature" but by an appreciation of nature in and of itself, as an organic whole. The art of this period moved beyond "a vision of nature shaped by human measures of time and by the dynamics of historical development." If this interpretation is correct, it may reflect the shift to viewing nature as "ideal" when it is in its "pristine" state, in which human contact and development are controlled or absent, a sentiment that would culminate in the national parks. Be that as it may, the vision of nature confirmed America's special role, reinforcing the sense that natural laws substituted for the historical laws of Europe.[14]

Visual images of the landscape had wide public circulation through prints and reproductions in national periodicals. Exhibitions also drew large paying audiences.[15] Art was quite possibly the primary prism through which Americans, in the North at least, symbolically comprehended the country's spatial dimension (including the link to nature) and historical role (its civilizing mission, for example). The landscape came to tell the story of America, its very being; in this way landscape art served as an "institutionalized aesthetic," a description of the cultural topography through time and space.[16]

The soil, the land, the blood of life, from which freedom grew—that land was the visual imagery, the painting, the poem that Lincoln would recite at Gettysburg. But in the carnage of the Civil War, the association of blood and soil was all too literal. The violence "on the ground" was conveyed by a Union soldier who wrote of a fellow prisoner at the Andersonville prisoner-of-war camp in Georgia: "One man . . . lies near our tent . . . debilitated with swelled feet from exposure to the sun or dropsy, chronic diarrhea, and neglect of cleanliness was found to have the lower part of the body near the rectum eaten into by maggots, which literally swarmed on him."[17] This was the picture of war; the American landscapes that confronted the populace were Shenandoah, Andersonville, Bull Run, and Gettysburg and the killed and maimed in the hundreds of thousands.

Nature—the land—would once again be called upon as mother earth, the fertile source of life, this time sacralized with the blood of its sons. In the death of soldiers, their intermingling with the soil and their supreme civic sacrifice, individual and regional differences would be transcended. The sacrifice of Union soldiers became a virtue in that it was the ultimate expression of dedication to the Republic. The shed blood would be symbolically transfigured as "cleansing" the sins of the past, above all the sin of slavery. In Lincoln's mind slavery was, after all, an *American* issue.

Lincoln and the "Mystic Chords" of the Land

The visual imagery of the land as the organic body, as the fertile soil that gave birth to the Republic and represented the nation in all its glory, had a strong impact on the public imagination, and it was the very imagery that held Lincoln. Indeed, the fear that slavery might rip the Union asunder was, implicitly at least, of greater concern to Lincoln than the institution of slavery itself, though he realized the mutual relationship of these issues.[18] And we can observe how Lincoln turned to the land to try to hold his nation together and, once war had broken out, to mend the jagged cut in the body of the Republic.

Lincoln's First Inaugural Address, on March 4, 1861, was delivered in dark circumstances. Jefferson Davis had been inaugurated as the president of the Confederacy two weeks earlier. Lincoln arrived in Washington under a cloak of secrecy and the protection of federal soldiers. In his speech, Lincoln appealed to the indivisibility of the land, the country, moving from its physical contiguity to national and moral commonality:

> Physically speaking, we cannot separate. We cannot remove our respective sections from each other, nor build an impassable wall between them. A husband and wife may be divorced, and go out of the presence, and beyond the reach of each other; but the different parts of our country cannot do this. They cannot but remain face to face; and intercourse, either amicable or hostile, must continue between them. Is it possible then to make that intercourse more advantageous or more satisfactory, after separation than before? Can aliens make treaties easier than friends can make laws? Can treaties be more faithfully enforced between aliens than laws can among friends? Suppose you go to war, you cannot fight always; and when, after much loss on both sides, and no gain on either, you cease fighting, the identical old questions, as to terms of intercourse, are again upon you.

Suggesting that Providence "has never yet forsaken this favored land," Lincoln turned directly to the separatists: "In *your* hands, my dissatisfied fellow countrymen, and not in mine, is the momentous issue of civil war." In the final paragraph, Lincoln issued a last appeal: "I am loath to close. We are not enemies, but friends. We must not be enemies. Though passion may have strained, it must not break our bonds of affection. *The mystic chords of memory, stretching from every battlefield, and patriot grave, to every living heart and heath-stone, all over this broad land, will yet swell the chorus of the Union, when again touched, as surely they will be, by the better angels of our nature.*"[19]

What is remembered most about this speech a century and half later are the iridescent phrases "the mystic chords of memory" and "the better angels of our nature." But what is equally, if not more, significant are the words between those phrases—words that in part anticipate a central thread of the Gettysburg Address. "Chords" today would be rendered as cords, or threads, and here Lincoln draws the association between land, the bodies of lost soldiers, and the nation: the "mystic chords of memory," the sense of nationhood that ties us all, "stretching from every battlefield, and patriot grave, to every living heart and heath-stone," evoking the connection between the land, nationhood, and civic sacrifice, "all over this broad land, will yet swell the chorus of the Union," the "chorus" representing the essentially unitary nature of the nation.[20]

In a remarkably direct way, Lincoln asserted the connection between land, nationhood, and government in his Second Annual Message to Congress. Furthermore, he stressed the land as the most important part of a nation and, in the case of the United States, adapted only for one "national family":

> A nation may be said to consist of its territory, its people, and its laws. *The territory is the only part which is of certain durability. "One generation passeth away, and another generation cometh, but the earth abideth forever." It is of the first importance to duly consider, and estimate, this ever-enduring part.* That portion of the earth's surface which is owned and inhabited by the people of the United States, is well adapted to be the home of one national family; and it is not well adapted for two, or more. Its vast extent, and its variety of climate and productions, are of advantage, in this age, for one people.

The land is the only durable part—it is the organic body of the body politic. Lincoln went on to quote his inaugural address, repeating verbatim the passage that begins, "Physically speaking we cannot separate" (quoted above), highlighting the significance he attached to that passage. "There is no line, straight or crooked, suitable for a national boundary, upon which to divide," he said, also suggesting that separation would result in onerous trade regulations on Americans in the "great interior." He added that, given the lay of the land, any (successful) separation would be temporary and reunion would eventually follow, but at great cost: "Our national strife springs not from our permanent part; not from the land we inhabit; not from our national homestead . . . In all its adaptations and aptitudes, it demands union, and abhors separation. In fact, it would, ere long, force reunion, however much of blood and treasure the separation might cost."[21]

The point about the physical union of the country was about not only the *extent* of the land but also what the land *signified.* In one centrally important respect, the land was of much greater symbolic importance to the Union than to the Confederacy. In the Confederacy, the land was a resource and basis for independence from the North, but belonging and rank in society rested on the fulcrum of kinship, or blood descent, primarily race. In the relative diminution of race and the absence of caste as a formal criterion of nationhood in the North, the land (and nature generally) took on a greater significance. The southern landscape was a hodgepodge of large and small slave plantations, small farms, and homesteads.[22] The North, in contrast, was more likely to be carefully and rationally surveyed and its lands demarcated (see the discussion of the Northwest Ordinance in Chap. 2). Carefully demarcating the national domain—the very process of rational, scientific cartography—was essential to defining and building the nation, and it was interconnected with attributing the birth of the nation to nature. Thus the Union promised not only the widest possible domain of the United States but also a bounded, homogeneous (a singular people), and "flat" (as in the absence of feudal hierarchy) land. The result of the Civil War clearly accelerated this process.[23] Lincoln alluded to this central difference between North and South—the role of kinship or blood descent in determining place and membership in society—in a message to a Special Session of Congress on Independence Day, 1861: "This is essentially a Peoples' contest. On the side of the Union it is a struggle for maintaining in the world, that form, and substance of government, whose leading object is, to elevate the condition of men—to lift artificial weights from all shoulders—to clear the paths of laudable pursuit for all—to afford all, an unfettered start, and a fair chance, in the race of life."[24]

No wonder Lincoln pointed, repeatedly, to the Declaration of Independence and to the assertion that it is "self-evident" that "all men are *created* equal" (emphasis added). Such rights of equality were a birthright and, as such, a product of nature, like the nation itself. Land, nature, the nation, man, rights—they were all of a piece.

Interestingly, Lincoln stressed this characteristic of the United States, as he promoted it, with regard to other "outsiders"—namely, immigrants. In a speech in Chicago on July 10, 1858, he referred to the generation of Americans of 1776 and Independence Day celebrations. Talking at first of the generation of 1776 he noted:

> We have beside these men—descended by blood from our ancestors—among us perhaps half our people who are not descendants at all of these men, they are men who have come from Europe—German, Irish, French, and Scandinavian . . . If they look back through this history to trace their connection with those days by blood, they find they have none, they cannot carry themselves back into that glorious epoch and make themselves feel that they are part of us, but when they look through that old Declaration of Independence they find that those old men say that "We hold these truths to be self-evident, that all men are created equal," then they feel that that moral sentiment taught in that day evidences their relation to those men, that it is the father of all moral principle in them, and that they have a right to claim it as though they were blood of the blood, and flesh of the flesh of the men who wrote that Declaration and so they are. That is the electric cord in that Declaration that links the hearts of patriotic and liberty-loving men together, that will link those as long the love of freedom exists in the minds of men throughout the world.

Lincoln went on to condemn arguments for slavery, noting that "they are arguments that kings made for enslaving the people in all ages of the world," suggesting that such arguments will not stop "with the Negro."[25] In a speech Lincoln made later that same year, following the second Lincoln-Douglas debate (one in a series of seven debates Lincoln had with the Democratic senator Stephen A. Douglas during the 1858 Illinois senatorial campaign, mostly on the slavery issue), he linked the abomination of slavery and the moral significance of the land: "The Republican party . . . hold[s] that this government was instituted to secure the blessings of freedom, and that slavery is an unqualified evil to the negro, to the white man, *to the soil,* and to the State."[26]

During the war Lincoln advocated trying to repatriate freed slaves to Africa or the tropics of the Americas. From the perspective of the present-day United States, this appears as something of a blemish on Lincoln's record. Lincoln made this proposal in the context of "solving" the slavery issue, but to him it made sense morally as well in the context of peoples tied to certain lands. The enslaved or their forefathers and foremothers had been rent from *their* (African) lands. (It is noteworthy that the "Back to Africa" sentiment reflected beliefs among some slaves and anticipated, in later generations, views of some black American movements.) Indeed, Lincoln compared the story of the slaves in America to the story of Exodus, drawing a parallel with the Hebrews enslaved in Egypt. In his eulogy in 1852 for Henry Clay, a senator known as the "Great Compromiser" in the sectional struggle, Lincoln approvingly quoted Clay himself: "There is moral fitness in the idea of returning to Africa her children, whose ancestors have been torn from her by ruthless hand of fraud and violence. Transplanted in a foreign land, they will carry back to their native soil the rich fruits of religion, civilization, law and liberty." Lincoln, noting that Clay had said this twenty-five years earlier, went on to state that "Pharaoh's country was cursed with plagues, and his hosts were drowned in the Red Sea for striving to retain a captive people who had already served them more than four hundred years. May like disasters never befall us!"[27] The last sentence strangely echoes Thomas Jefferson's reflection on slavery when he wrote, "I tremble for my country when I reflect that God is just; that His justice cannot sleep for ever: that considering numbers, nature and natural means only, a revolution of the wheel of fortune, an exchange of situation, is among possible events: that it may become probable by supernatural interference!"

Ideas of reparation would, of course, fall by the wayside. The enslaved would be freed and given a fair(er) chance in the "race of life": the long, arduous, meandering path to full civic participation of African Americans could begin. "Belonging" came from the land, not blood. Blood, instead, reflected sacrifice and freedom. At Gettysburg, the land would be configured to shape a reborn nation.

To Make a Nation

At Gettysburg Lincoln took the mood of the organic, bodily place of the land, so resonant in the American heart, and transformed it. By transform, he did not diminish that connection; on the contrary, he took that sense of the land to transcendent heights. It was a national relation to the soil, it was the *nation* that felt the touch, the scent, and, above all, the sight of that rich earth. The farmer and the homestead were not replaced but were transcended by broad vistas of nature, parks, and battlefields and by the stories of collective struggles and journeys.[28] To achieve such a transcendent act, Lincoln had to appeal to the wellspring of America's imagination of itself, its tie to nature, and paint it on a broader and grander canvas. And in doing this, he would bring out the apparently inexorable connection between the soil, nationhood, and republican, civic politics:

> Fourscore and seven years ago, our fathers
> brought forth on this continent a new nation,
> conceived in liberty and dedicated to the proposition
> that all men are created equal.
> Now we are engaged in a great civil war,
> testing whether that nation,
> or any nation so conceived and so dedicated,
> can long endure.
> We are met on a great battlefield of that war.
> We have come to dedicate a portion of that field,
> as a final resting place for those who here gave their lives
> that that nation might live.
> It is altogether fitting and proper that we should do this.
> But, in a larger sense, we cannot dedicate,
> we cannot consecrate, we cannot hallow
> this ground.
> The brave men, living and dead, who struggled here,
> have consecrated it far above our poor power to add or detract.
> The world will little note, nor long remember, what we say here,
> but it can never forget what they did here.
> It is for the living, rather,
> to be dedicated to the unfinished work
> which they who fought here have thus far so nobly advanced.
> It is rather for us to be here dedicated
> to the great task remaining before us—
> that from these honored dead we take increased devotion
> to that cause for which they gave the last full measure of devotion—
> that we here highly resolve that these dead shall not have died in vain,

that this nation, under God,
shall have a new birth of freedom,
and that government of the people, by the people, for the people
shall not perish from the earth.[29]

Poetry elicits a felt experience and a picture, a temporal rhythm and a spatial form. It is a complete, enclosed, yet transcendent experience. At Gettysburg Lincoln, who wrote poetry, made of America a poem. Scholars sometimes like to talk of national "narratives"—stories nations like to tell about themselves and their founding myths. In contrast to the linear quality of the narrative, poetry transcends the passing moment—it transfigures the here and now, whenever and wherever that here and now is, into a moment made eternal. In that sense, the moment is sacralized in contrast to other, mundane moments, which are quickly forgotten. Venerated monuments have the same effect, on a larger scale. Through the address, Gettysburg became both poem and monument. At Gettysburg Lincoln incorporated and transformed America's sense of itself, of its place, to the point where it was self-evident and presumed. This was the effect of his address.

The address also reflected a sentiment found in another poem, of another nation from another war, Rupert Brooke's "The Soldier" (quoted in the introduction). Through the fallen bodies of its soldiers, the nation and the soil are one; physical body, body politic, nation, and land become fused. In death and sacrifice, the nation's immortality is revealed. These themes resonate throughout Lincoln's address. One cannot but wonder if Lincoln, an avid reader of Shakespeare, felt the sentiments and thought the thoughts, perhaps even could hear the words, of Henry V addressing his troops before the Battle of Agincourt as the president composed his own dedication: "If we are mark'd to die, we are now / To do our country loss . . . / From this day to the ending of the world / But we in it shall be remembered / We few, we happy few, we band of brothers; / For he today that sheds his blood with me / Shall be my brother."

Lincoln deeply felt this mood of kinship with the soil on a personal level. This is revealed in a verse of Lincoln's own poetry. In a poem titled "My Childhood—Home I See Again" he mused: "The very spot where grew the bread / That formed my bones, I see / How strange, old field, on thee to tread, / And feel I'm part of thee!"[30]

Lincoln evoked, as we noted earlier, the land as source of birth and national life—"conceived," "brought forth," and "new birth." The death of the soldiers is transcended through the immortality of the nation: the soldiers "gave their lives so that the nation may live." The land, then, "constitutes" the people. But Lincoln was careful to weave in that the Republic was of the soil, as it was of the people, and the people could be defined only through civic politics, just as the land constituted them. In other words, land, people, and republican polity were inextricably whole. Thus this new nation was "conceived *in liberty*"; the very act of conception, the creation of the nation, was rooted in "natural" principles—this nation at birth was "dedicated to the proposition that *all men are created equal*"; the "new birth" of the nation following the Civil War is a new birth of "freedom"; and, finally, that "government of the people, by the people, for the people *shall not perish from the earth.*" "Earth" here can be construed in two senses, both as simply the place where we live and, more redolently, as the soil of life. One can even read into this reference to the earth, perhaps, an allusion to the biblical phrase Lincoln did use elsewhere, that the "earth abideth forever."

The symbolic expression of violence (as a virtue or as a national tragedy) through military cemeteries, monuments, and commemorations has thus been at the very center (literally and metaphorically) of national self-representation, and this was true as the United States sought not just "union" but nationhood. The new, unified image of the nation is also conveyed through the symbolic rendering of violence as the ultimate form of civic sacrifice, whereby lives are *given* (not simply "lost") to the nation and its cause. The historian Garry Wills notes how Lincoln in his speech never refers to himself (as in "I"); he uses only the national "we." Nor does he refer to the dead by name; he speaks only of "what they did here" or "these dead". "The general or generalizing article—*a* great civil war, *a* great battle," folds a particular struggle into a larger national saga and purpose. As Wills notes, "The draining of particulars from the scene raises it to the [ideal] of a type." The prose form shifted the mourning from the individual and the family to the "larger community's sense of purpose."[31] So a unitary civic and national community is *embodied*, created, as it were, a posteriori—the body politic is formed and commemorated at the same time, as the country mourns its dead. The blood spilt cleanses the sins of the past—this is implicit in Lincoln's use of the term "reborn nation" but also stated in explicit terms by others. On the eve of the Civil War, James A. Garfield, then a representative in the Ohio legislature (he later served in the Union army and was briefly the president of the United States in 1881 until he was assassinated), wrote that the Bible taught that "without the shedding of blood there is no remission of sins."[32] National, collective mourning drew the nation together in the experience of death.

The Union made a nation through transposing individual sorrow and grief into a story of collective redemption. This is evident in Lincoln's well-known letter in late 1864 to Mrs. Lydia Bixby. She was the mother of five sons, all of whom were reported to have died gloriously in battle for the Union. "I pray that our Heavenly Father," Lincoln wrote, "may assuage the anguish of your bereavement, and leave you only the cherished memory of the loved and lost, and the solemn pride that must be yours, to have laid so costly a sacrifice upon the altar of Freedom."[33] In the war song "Battle Hymn of the Republic," essentially the anthem of the North and sung by Union troops marching to the front, the same merging of the individual into the nation is expressed in the final verse:

In the beauty of the lillies Christ was born across the sea,
With glory in His bosom that transfigures you and me:
As He died to make men holy, let us die to make men free,
While God is marching on.[34]

The symbolism of violence was not only about the sanctification of the soil through the blood of lost soldiers, the transcendence of individual grief in a national cause, or the cleansing of the sins of the past. The monopolization of violence by central government—where the only *virtuous* violence was that of the state representing the common weal—created a "civilized space" where citizens (as *civilians*) could shape their society without fear for their personal safety.[35] The monopolization of violence was critical for creating, by extension, a common citizenship. As noted earlier, up to the Civil War the courts could never determine which citizenship superseded the other—that of the individual states, say Virginia, or that of the United States.[36] With the Fourteenth Amendment citizenship was common to black Americans, as well. The military has traditionally been glorified as the ultimate expression of independence and self-determination. By destroying the Confederate armies, the Union forces were, paradoxically (given the massive destruction involved), creating a common civic space. "Virtuous violence" would now be held by one, federal authority, and this was of symbolic as well as political value. The representation of violence in military cemeteries and monuments would reflect this.

The centrality of citizenship, and the link to the land, were accentuated further in the nature of the sacrifice. The blood lost was for the "cause" of "freedom" and of "government of the people." The theme of sacrifice, blood, and loss has often been invoked in nationalist causes, in the extreme case glorifying a *Volk,* a people defined through blood descent. The perigee of such nationalist expression is the xenophobic and chauvinistic nationalisms propagating their own racial superiority. In the case of the Union, and especially Lincoln, it was the opposite: the "cause" (for which blood was shed) was of universal import, independent of blood descent. Lincoln commented how the United States carried a universal, worldwide burden in trying to protect freedom.

The threading of body and soil, and civic sacrifice and nationhood, would culminate in the establishment for the first time of official military cemeteries that commemorated the war dead. Prior to this, war dead were buried in civilian cemeteries, if they were buried at all. Military cemeteries were initiated by an act of Congress of July 17, 1862.[37] The timing of the act is significant: military cemeteries and monuments became powerful—central—mediators, or symbols, for "marking" nationhood.

National Monuments and the Mediation of Nationhood

War monuments and cemeteries evoked a mystical sense of experiencing the transcendent. The military monuments and cemeteries represented death, the "other side," and national rebirth. But in another, critical respect, the flourishing of war monuments and carefully tended public military cemeteries in the wake of the Civil War reflected a dramatic switch for the then deeply and overwhelmingly Protestant United States. The Protestant reformers had rejected any "mediation" of the believer's relationship with the Almighty. This, after all, was at the crux of the Protestant Reformation. The mediating role of the Catholic Church—through the church hierarchy and priesthood, or through icons of any kind (such as religious images of the Virgin Mary and of the saints)—was swept away. The believer experienced God without intercession. Particularly in the case of denominations and sects such as the Baptists and Methodists, God was a palpable and direct individual experience (such that nineteenth-century revival meetings created frenzies of emotional ecstasy, as the Spirit moved them, causing Quakers to "quake" and Shakers to "shake"). The Catholic Church hierarchy broken, the experience of God individualized and denominationalized, Protestantism spawned a veritable embarrassment of riches—not only in terms of the Protestant's role in the rise of capitalism but in the multitude of sects, denominations, and churches that arose in place of the one universal Church of Rome. The rejection of the mediative role of the Catholic Church led the reformers also to demur at the pilgrimage, so characteristic of medieval life. Pilgrimage expressed the idea that God could be "experienced" only at certain points of mediation, often through icons of some kind—statues of the Virgin Mary and relics of miraculous sites, for example—in places such as Rome and Jerusalem.

In this context, the rise of national monuments and parks that were mediating symbols of nationhood, in the wake of the Civil War and ever since, is a remarkable switch in American history (although it was the nation that was mediated and not, at least directly, God).[38] Protestantism had fostered the direct relationship with God, and nature was an expression of God. The Jefferson yeoman farmer was directly in contact with nature, which in turn led to a partiality to the "local" and so reflected the Protestant sentiment of the country (and its many sects and denominations). But this created the potential for fission, just as the Protestant rejection of Catholic Church mediation had let a thousand churches rise in its place. That potential was realized in the secession of southern states. The creation of a unitary republic demanded unifying symbols. A Jeffersonian, face-to-face community could not be built in a country of tens of millions; the national sense of community had to be mediated and anchored. The American imagination—the "likeness of America" that was to be shared in the mind of the public and promoted through civic education—was not the yeoman farmer or the frontiersman. That practice created the possibility of a thousand republics—a man's (or, we can now add, a woman's) homestead could be his or her republic. Gettysburg, national parks, the Statue of Liberty, and other such monuments and parks told a unifying story, a national story, a republican story, and a story of historical destiny; a story could be imagined, in the mind's "I," only in national terms.[39] It is not suggested here that these symbols were created out of some conscious desire to "manipulate" national identity; rather, they followed the premise of combining "Union *and* liberty," in Daniel Webster's words. Thus, at Gettysburg Lincoln would seek symbols and ideas to bind when trying to unite a country on common republican principles in which he deeply believed.

Thus, following the Civil War, and with the receding frontier and growing urbanization, we see the flurry of the founding of monuments and memorials and the establishment of national parks. The parks, which mediated the tie to nature for an increasingly urban citizenry, themselves incorporated national themes about place and destiny. And with the monuments and parks a new form of pilgrimage arises—in the form of tourism to these sites. Tourism, as a mass phenomenon, began after the Civil War, which may be attributed in part to easier transportation, through the railroads, for example; but the destinations of such tourists—namely, historical sites and parks—reveal the desire to "experience" nature, to experience "America."[40]

War monuments and cemeteries become the cardinal points of America's new civil religion. War is the ultimate political act, the final and absolute expression of a people's commitment to worldly existence, to the land. It is an act of transcendence of the self into an epic story, a story that brings the individual into the drama of nations, of the history of good and evil. Therefore, not only body and soil but self and nation, now and forever, the here and now and the All in All, become fused in the frisson of the "good war." Contrast the poetry of Rupert Brooke with the poetry of disillusion with war, such as the work of another World War I poet, Wilfred Owen. In innocence lost, war is painful and alien and, above all, cast as a lonely experience. Death in combat is not a glorious flight into immortality. Death bespeaks loneliness, not in the verdant soil of the nation but drowning in one's own blood, in the mud, and in the stench of rotting bodies:

> If you could hear, at every jolt, the blood
> Come gargling from the froth-corrupted lungs,
> Obscene as cancer, bitter as the cud
> Of vile, incurable sores on innocent tongues—
> My friend, you would not tell . . .
> To children ardent for some desperate glory,
> The old Lie: *Dulce et decorum est*
> *Pro patria mori.*

"How sweet and becoming it is to die for one's country."

Monuments, Cemeteries, Parks, Urban Design, and the Public Square

The monuments, cemeteries, and parks commemorating "virtuous violence"—to fight for one's country, to make the final sacrifice—are central markers of civic life in post–Civil War America. If the Northwest Ordinance was critical in etching a certain Jeffersonian ideal into the landscape, making it a moral landscape in every sense of the term, war cemeteries and monuments would anchor the public life of a now unitary but also increasingly urban United States. If the picture or image of the common civic life of the early Republic was the independent farmer on his homestead, the aesthetic, and heart, of democratic and urban America was the "public square," often a park or garden of some kind—with a war monument at the very center of that square. And in the most sacred of public spaces, in Washington, D.C., where the Republic as nation and polity is most keenly felt, the remembrance of the war dead is pervasive, the very soul of the city and, by extension, of the Republic itself. Washington, D.C., as the sacred shrine to the nation and its dead is largely a post–Civil War development. The Arlington war cemetery itself began as a cemetery and shrine for Union dead—on land requisitioned from Robert E. Lee's family.

Indeed, it is in the burial and remembrance of the dead, and the relationship to death, that one can discern the contours of civic life of America. And the very patterns of urban design in the United States were significantly shaped by memorialization of the nation's military dead—especially in connecting urban America to the soil, the land of the nation. "The solitary or hidden grave is perhaps symbolic of an obligation finally discharged; the public graveyard is a reminder of duties constantly recurring; in the true meaning of the word, it is a monument, a 'bringing to mind,'" wrote the geographer John Brinckerhoff Jackson.[41]

In colonial America the most common form of burial was in the churchyard or in a cemetery next to the church. The graves were not located in an especially "natural" setting; well-tended lawns and trees and a carefully nurtured rural setting in cemeteries would appear only in the nineteenth century. The desire to contrive, as it were, the connection with nature would arise in an urban environment. Headstones and the graveyard itself, located in the center and unadorned and unconcealed, in colonial villages served to remind the populace of the fragility of life, a reminder to live a virtuous life of faith.[42] Indeed, Puritan headstones were typically illustrated with the skull and crossbones.

With the demise of the Puritans in the eighteenth and nineteenth centuries and the rise of more optimistic (salvation was in the reach of all who chose a godly path) Protestant denominations, especially Baptists, the way of death changed. For one thing, there was no state church but a multitude of denominations; the house of faith was largely a personal matter. Burial was taken out of the public center, both literally and symbolically. Headstone illustrations also changed, portraying cherubs and angels, expressing a certainty of salvation which the Puritans, with their doctrine of predestination (which suggested that hell was quite possible even for the fastidious believer), could not share.[43] It was in the unification of the country and the Civil War that death and burial became public once more, literally monuments placed in the public square, but this time not of the populace as a whole but of the nation's fallen soldiers. Thus, insofar as death is *publicly* commemorated, the duties that are "brought to mind" are of the *patria,* symbolizing and binding together the nation as a whole.

But the new patterns of burial, peripheral to the civic center of the towns and villages, did selectively influence the landscape architecture of the military cemeteries and monuments that followed more than half a century later. Interestingly, it was an urban designer, James Hillhouse, who first designed in the late 1790s a cemetery outside the city center of New Haven, Connecticut, the "New Burying Ground." The city fathers had wanted to beautify the city green, where the dead had been previously buried and whose remains would be reinterred in the new cemetery. Hillhouse's design—a grid pattern of square plots divided into smaller squares that could be "occupied" by families—became the standard for most American cemeteries. The grid design was the signature and organization of American republicanism—egalitarian, rational, bounded, and scientific—characteristic of the surveys that demarcated the landscape (as in the Northwest Ordinance) and now marked graveyards. The cemeteries were nonpublic and made allowance for family, as opposed to individual, plots, and "customized" markers. These markers commonly exceeded the modest headstones and were monumental in design, with urns, obelisks, columns, and statues to commemorate the deceased. No longer a "public reminder," these graveyards were, if at all possible, built in secluded places. Death was now a private, family matter.[44]

Mount Auburn Cemetery, established in 1831 outside Boston, was less influential than the Hillhouse design in terms of numbers of cemeteries that emulated it, but it had a fundamental impact on monuments and landscape architecture. It too sought to be secluded and private, but unlike Hillhouse's symmetry, Mount Auburn's appeal was in the romantic tradition, fitting in the gentle, informal contours of nature in a woodland setting. Graves belonged in nature, "dust to dust," or, as in the words of a clergyman at the Mount Auburn inauguration, "The child of nature is clasped again to the sweet bosom of its mother, to be again incorporated in her substance."[45] The setting as much as the graves inspires. As in the Hillhouse-style cemeteries, grief and mourning are private, out of the public eye—that is, until Gettysburg and the rise of official, military cemeteries and monuments.

The Mount Auburn model of resurrection-in-nature is emulated at Gettysburg and other later military cemeteries, but in an intensely public way. The pain of loss is public and collective and as such not only unifies but also creates a peculiar frisson, or excitement, as the mourner feels him- or herself soaring above in a national story that transcends time and space. Edward Everett, the great nineteenth-century public speaker who gave a two-hour oration at Gettysburg (the president's address, which followed, constituted "dedicatory remarks"), expressed this quality of public grief. He proclaimed in his oration that "You feel, though the occasion is mournful, that *it is good to be here.*" And then Everett added, "You feel . . . a new bond of union," and talking to the lost soldiers, "no lapse of time, no distance of space, shall cause you to be forgotten."[46] Gettysburg is the archetypal model of a military cemetery that is a "reminder of duties constantly recurring," and "in the true" and literal meaning of the word, "it is a monument, a 'bringing to mind,'" which unites the sacrifices of a past generation with those living today.

Everett, and more broadly the rural cemetery movement that was inspired by Mount Auburn (and, in turn, was a source of inspiration for the Gettysberg cemetery and monument), sought, as Garry Wills writes, to *prolong* the grief. Rural cemeteries were designed to foster the contemplation of nature, the betwixt-and-between of life and death. They sought to capture what

Rousseau had described experiencing, the loss of self in the greater whole, in the communion with nature.[47] The liminal state *is* that state of *unio mystica,* in which mere language and intellectual categories just obscure the transcendent totality, the All in All—"To be uplifted to the clarity of ecstasy, to wander on the solitary heights of contemplation stripped of forms and images, tasting union with the only and absolute principle."[48] Even the "self" as a discrete "being" detracted from that sense of being at one with nature. And this underlay the desire to prolong the national grief and to commemorate it. National mourning dissolved the self in a greater union, in nature's nation. National sadness is nurtured, through monuments and national commemorations. National mourning is a transcendent experience, transcending time and space. The mourner of a personal loss wishes ultimately to recover from private grief—and it is expected that, after a certain period of time, such a mourner will resume a relatively routine life. Death in the name of the Republic is "sweet and becoming."

Everett in his oration at Gettysburg observed how he hesitated to raise his "poor voice to break the eloquent silence of God and Nature" (silence being more sacred than simple words) and that he was ready to take off his shoes "as one that stands on holy ground" (to be in palpable touch with nature). He spoke of the death of the soldiers: "As my eye ranges over the fields whose sods were so lately moistened by the blood of gallant and loyal men, I feel, as never before, how truly it was said of old, that it is sweet and becoming to die for one's country. I feel as never before, how justly, from the dawn of history to the present time, men have paid homage of their gratitude and admiration to the memory of those who nobly sacrificed their lives, that their fellow men may live in safety and in honor." War was not glorified; "next to defeat, the saddest thing was victory." It was the mourning that was played upon. Later in his oration, Everett recounts a dying soldier's last wish: "'Tell my little sister not to grieve for me; I am willing to die for my country.' . . . When, since Aaron stood between the living and the dead, was there ever so gracious ministry as this?"[49] At the end of his speech, he tells his audience: "You now feel it a new bond of union, that they shall lie side by side, till the clarion, louder than that which marshalled them to combat, shall awake their slumbers . . . [As] we bid farewell to the dust of these martyr-heroes, that wheresoever throughout the civilized world the accounts of great warfare are read, and down to the latest period of recorded time, in the glorious annals of our common country, there be no brighter page than that which relates The Battle of Gettysburg."

National melancholy lingers. Monuments eternalize, or seek to eternalize, the sadness of the sacrifice and in so doing sacralize the cause. The civic center, figuratively and literally, of urbanizing America was, as we noted, the public square, often in a "natural" form. The monument anchors this square and hence the city itself. Those monuments were the epicenter of urban planning at the turn of the century and of the City Beautiful movement. Those monuments, at their most venerated and particularly in most sacred center of them all, Washington, D.C., are the "holy of holies," icons where the nation celebrates (some would say worships) itself—indeed, has to in order to be a nation at all.

Thomas Jefferson took an interest in urban design, despite his trepidation about the city, and made suggestions for the design for what became Washington, D.C. Jefferson proposed a grid system, an extension of (or as a fragment of) the grid of the national survey associated with the Northwest Ordinance. A city made up of unvarying uniform blocks with a public square at its center, with each block made up of individual landholdings, has indeed served as the model for the great majority of American cities. The urban grid, like the rural grid, would supposedly promote equality, independence, and virtuous citizenship.[50] Jefferson precisely described his vision of Washington: his plan called for blocks of six hundred square feet, all at "right angles," organized in a grid pattern, and eleven blocks wide on an east-west axis and three blocks deep. He envisaged a mall or "public walks" linking the Capitol and the presidential residence. The grid system, he believed, would allow for orderly expansion of the city. It was a striking expression of Jeffersonian democracy and belief in rational progress.

If such a Jeffersonian vision of urban design succeeded in most of the country, it was struck down quickly and forcefully by Pierre Charles L'Enfant, the designer of Washington, D.C., who wanted a city of monumental grandeur. L'Enfant responded by saying that the grid plan was "tiresome and insipid." He called for major, monumental buildings, open spaces, and open public areas. Jefferson failed and L'Enfant prevailed in their respective visions of the "Federal City." Washington, D.C., would become the monumental center of the country but still as the capital of a confederation of states. Significantly, L'Enfant's design called for monuments, designating fifteen squares (this being

1792) for, he wrote, "statues, Columns, Obelisks, or any other such ornaments *as the different States may choose to erect.*"[51] The national monuments and extensive landscaped parks would be erected mostly after the Civil War, as the city became a truly national capital and symbol.

Jefferson may have been averse to the city, but he did not promote nature as something that should be pristine and untouched; he believed in "rural society," not "rural solitude." Humankind lived in a symbiotic relationship with nature, and that included tilling the soil and other forms of rural industry.[52] By the post–Civil War period, the perspective of nature had changed. The monuments, parks, and other icons that were to mediate nationhood had to be sacred, without utilitarian purpose, and preserved. Similarly, with the "end of the frontier," parts of nature had to be protected and conserved if humankind's essential tie with the natural world was to be maintained. Instead of the "Middle Landscape," where industry and nature coexisted, capitalism and industry were viewed to be opposition to the environment. In this context, the romantic ideal—with "natural" twisting forms and growth—of Mount Auburn and of parks and monuments became more popular.

Cities sought to develop and maintain green public areas; Frederick Law Olmsted's design of Central Park in New York City was the most famous. The creation of such large areas as public parks was without precedent; the great gardens, such as Kew in England, were royal preserves or, like Gramercy Park in New York, private.[53] The idea remained that these urban nature preserves would have a "civilizing" effect and were thus public squares in multiple senses. "It is one of the ironies of our history that the romantic environment remained an urban and suburban phenomenon," wrote J. B. Jackson. "Only in the 1870s," he noted, "did it become possible to interest the American public in the wisdom of preserving the scenic wonders of the wilderness."[54] Washington, D.C., was at the center of all this, but even the other, "tiresome and insipid" gridlike cities sought green civic parks and public squares, monuments at the center, with the city radiating outward from this axis—if not an *axis-mundi,* an *axis-patri.*

By the last decade of the nineteenth century, an intensive effort was begun to develop civic centers in American cities, with fountains, ornamental benches, statues, and the like under the aegis of the City Beautiful movement. War memorials, located in civic meeting points, became focal points in city design just as there was increasing salience of memorialization in Washington, D.C. Indeed, sometimes it took war memorials or the appellation war memorial to build all kinds of public facilities.[55] Such public squares and memorials became the "hubs" of the cities, physically and symbolically the heart of city life; civic life was tied to the symbols of sacrifice and nature and, by extension, to the nation itself. "A hierarchy of shrines" is evident, beginning in Washington, D.C., and continuing to other cities, but often also within cities themselves. The major monuments are the epicenter of public life, while other monuments—concerning lesser-known figures, for example—are more peripheral.[56] Monuments girded the city spatially, in much the same way that national holidays (in which war commemorations figure prominently as well, from Veterans Day to Memorial Day) gird the flows of the calendar and time: they are sacred points of reference that locate day-to-day life in the larger sweep of time and space, bracketing the mundane in a larger national epic.[57]

Spatial Rhythms: Changing the Past

FOUR

If he is to live, man must possess and from time to time employ the strength to break up and dissolve a part of the past: he does this by bringing it before the tribunal, scrupulously examining it and finally condemning it . . . [Replacing this past] is an attempt to give oneself . . . a past in which one would like to originate in opposition to that in which one did originate.

—FRIEDRICH NIETZSCHE

PEOPLE LOCATE THEMSELVES historically and spatially to make the present transcend itself instead of existing only as a transient, quickly forgotten moment. The present becomes meaningful only if it is extended "transhistorically," that, is by eternalizing the moment. But there is a certain rhythm, or dialectic, in monumentalism. In the very act of seeking to "objectify" a point in time, the larger, transcendent purpose may be diluted. It is like a famous painting, say the Mona Lisa, pictured over and over again; its ethereal quality is progressively lost. And in objectively—monumentally—presenting a moment, the monument itself risks becoming fixed in time and place. As the Dutch historian Johan Huizinga wrote in his book *The Waning of the Middle Ages*: "Every thought seeks expression in an image, but in this image it solidifies and becomes rigid. By this tendency to embodiment in visible forms all holy concepts are constantly exposed to the danger of hardening into mere externalism. For in assuming a definite figurative shape thought loses its ethereal and vague qualities, and pious feeling is apt to resolve itself in the image." Writing of Catholicism on the eve of the Reformation, Huizinga further noted that "piety had depleted itself in the image, the legend, the office. All its contents had been so completely expressed that mystic awe had evaporated."[1]

The nation (or the church in Huizinga's description) is embodied—figuratively substantiated—in those symbols. As the Reformation turned from quarrel to combat in the mid-1500s, incipient Protestant groups swept through Catholic churches destroying the images, symbols, and artifacts of the church. This physical destruction also reflected a rejection of the mediative relationship of the church with God. The Protestant churches that replaced the Catholic churches were spare and austere, absent of the ornamentation of the Catholics. Simple crosses (if that) would replace Catholicism's crucifixes, with the figure of Jesus nailed to the cross. The cult of the saints and the Virgin Mary, and their figurations, disappeared altogether.

Symbols of nationhood may thus "harden," generating a certain nonchalance or even becoming almost invisible.[2] Rather than monuments of time eternal they represent the "dead hand" of the past. Those symbols thus become the targets of rebellious, reformist, and even revolutionary change. Because such symbols and monuments arrange "place," locating and orienting peoples spatially and temporally, and are critical in binding and mediating the body politic, they lie at the heart of such contentions; they determine who "belongs" to the nation and on what terms (for example, through assimilation). These conflicts are about the organization of public space—socially, morally, physically, and politically. These symbols—monuments, national parks, and historic sites—become the object of destruction (as in the Reformation or in Eastern Europe after the collapse of communism) or, only somewhat less dramatically, reinterpretation. The destruction of statues of Marx and Lenin, as well as the myriad other symbolic changes, from renaming streets to establishing museums of liberation to the collapse of the Berlin Wall, were the first acts of newly liberated countries in Eastern Europe, just as the physical destruction of Catholic images, statues, and icons was the first expression of mass and open revolt in the Reformation. "It is precisely this demand that greatness shall be everlasting," Nietzsche observed, "that sparks off the most fearful of struggles."[3]

In the United States such public symbols are reinterpreted and supplemented with other icons and markers. Indeed, in the past three to four decades, changes in the moral landscape of monuments, museums, parks, and other sites suggest an almost sweeping change in the meaning of "place" in America. This is especially evident in the changing representation of violence. The dynamics that lead to such questioning and reinterpretation do not concern us here. Why exactly and at what point did, for example, the "Founding Fathers" of freedom and republicanism become, for some, simply white, propertied males with at base venal and narrow interests (determined by their race, class, and gender) is an issue of great interest in intellectual and political history, but one we put aside here. Rather, we seek to elicit the process whereby the inextricable—the tie between the people and the land—became extricable and, as such, came to be embodied in the symbols of the nation. What emerges is a conception that many cultures, and peoples, each with its own "moral spaces," can (or should) coexist in one fixed geography; that these cultures can have varying relationships "to the land" and even consider their home as elsewhere. This has fundamental implications for the character of public space, citizenship, the rules of politics (including reinventing exactly what is an acceptable "culture"), and defining who "we" are as a nation and as individuals.

| *Parks and Monuments: The Changing American Landscape*

Almost all federal parks and monuments were established after the Civil War, even all the federal parks and sites dedicated to or associated with the Revolutionary War of 1776 (beginning with the Washington Monument in 1885). Yellowstone National Park, which was established in 1872, is usually credited as the first national park. However, in 1864 Lincoln signed a bill creating a state preserve through a combined federal-state effort when land was transferred to the state of California: the (in the language of the bill) "'gorge' in the granite peak of the Sierra Nevada Mountains—known as Yo-Semite alley," was designated for "public use, resort, and recreation."[4] Other historic sites, now under the National Park Service, include the White House, which is dated as a federal site from 1817. The National Park Service itself was established in 1915 in order to bring the management of all the parks, monuments, and historic sites under a single administrative umbrella.

Considering the approximately four hundred parks, monuments, and sites as a whole, what is conspicuous is the salience of military parks and monuments, especially in Washington, D.C. About seventy parks and sites are related to military events of one kind or another. The prominence of the military theme is even more marked early on; by the beginning of the twentieth century more than half the parks now under the aegis of the National Park Service were related to the military (almost all involving the Civil War itself). The number of parks was limited before the turn of the twentieth century—some fourteen parks and sites—but they are almost all major parks, such as Gettysburg (which was formally recognized as a national park in 1895), Shiloh, Arlington Cemetery, Yellowstone, Yosemite, Antietam, the Washington Monument, the Custer national cemetery (later to be called a national monument), and Vicksburg. The commemoration of war is pervasive in the landscape beyond monuments—street names, public buildings, and national holidays underscore the place of national military struggle in civic life.[5]

From roughly the 1970s a different pattern emerges, equally arresting. A series of parks and monuments were established, or older parks were redefined, as "ethnic heritage" parks (African American, Alaska Native, American Indian, Asian American, Hispanic, and Pacific Islander), as "human rights" parks (such as the Brown vs. Board of Education National Historical Site); or as women's history or women's rights parks (for example, Women's Rights National Historic Park, which celebrates the First Women's Rights Convention). Other parks express national atonement (such as Manzanar, which was an internment camp for Japanese, citizens as well as noncitizens, during World War II).[6] A number of parks (such as the Grand Canyon) were designated World Heritage Sites, under the auspices of the United Nations. More broadly, parks and monuments of all kinds offer "interpretations" (the National Park Service viewing interpretation as a critical mandate) that have moved away from the "heroic" description of American history and nature to a more "multiple perspectives" or multicultural approach. Violence itself came to be interpreted in vastly different ways, particularly at two highly visited sites in Washington, D.C., the Vietnam Veterans Memorial and the United States Holocaust Memorial Museum.[7]

Mapping the Land, Mapping History

Icons do not claim as such to be that which they represent. Nevertheless, as Albert Boime writes, the "icon's self-containment and its inevitable claim to direct spiritual status all but obliterates the distinction between itself and the exalted object of adoration." The icon in a palpable sense *becomes* the nation. Consider the evocative power of the flag to mobilize for battle or the intense passions that arise (for friend and foe alike) over questions of flag burning. Similarly, an art exhibit that "required" spectators to desecrate a flag by standing on it and displayed a flag draped in a toilet bowl generated a nationwide controversy.[8] Similar outrage would arise if someone took a hammer to a revered site such as the Lincoln Memorial. It is as if the icon, be it the flag or a monument, is the nation itself—and in a sense it is. Such icons demarcate the territorial and ideological domain of the nation (literally in the case of the flag); they also mark its place in history. In the case of the United States, what is celebrated about the past and past events is their promise for the future, the expansion of democracy, human rights, free labor, and liberty in its different forms.

But the contradictions, or at least paradoxes, contained in the icon are multifold. It is a material, inert object infused with values and beliefs; it is fixed in time and place but endowed with transcendent meaning; and, in the American case, it manifests the tensions implicit in the Declaration of Independence itself. This creed promotes universal values that apply to all humankind and should be promoted as such. But such creedal exuberance can be (and often is) taken as imperious, inimical to presumed American virtues such as national or individual self-determination (a tension felt internally in the United States as well as externally). Part of this tension results from the fact that while internally these beliefs may be seen as having universal validity—all, regardless of place or time of birth, are born with certain "natural" rights—externally those values may be seen as idiosyncratic, that is, representative of a certain time and place. Worse perhaps, they may be viewed as values of a particular, self-interested group of people—whites, males, the wealthy, Christians, and so on.

A case in point: in the nineteenth century liberal missionaries (especially Baptists and Methodists) were seeking to save the heathen in Africa (for a universal and monotheistic deity); they were also fighting slavery and seeking to establish basic rights for indigenous peoples. Conservative churches (such as the Church of England or the Dutch Reform Church) were not interested, or barely interested, in having nonwhites in their churches. Today, however, such liberal missionaries are viewed as tainted for destroying indigenous culture through proselytizing. The idea of assimilation, once a liberal policy of inclusion, is now suffering similar opprobrium in a more multicultural environment.[9] These dialectical qualities and the prying apart—or, in more contemporary parlance, the deconstructing—of the interplay of universal and particular elements of icons, especially national monuments and sites, effervesce in a more self-consciously multicultural America.

Shifting Grounds

Consider the metaphorical uses of "ground"—to hold one's ground; grounds for complaint; to give ground; to break new ground; to shift ground; forbidden ground; firm ground; to ground an argument: it is as if our lives are contoured by the landscape around us and our peregrinations through life are a series of spatial maneuvers. The physical and metaphysical qualities of "ground" become interlaced such that humanity's very essence is comprehended. Ground is the space upon which a nation (the Americans or the Lakota Sioux, for example) stakes its position. To lose ground implies decline—in moral as well as physical terms. The ground is the soil of the earth, and ground in this sense was for some mystics the essence of the soul. "The divine Ground of all existence is a spiritual Absolute, ineffable [but] susceptible . . . of being directly experienced."[10] "Being is posited as *Existence,* and the Mediating agency of this Being as *the Ground.*"[11] In the most prosaic sense, the ground is the surface upon which humans move or are at rest, and yet the ground is the physical axis—the coordinates—that locates people in a metaphysical universe, where matter and antimatter interpenetrate, fuse, and transport each other.

When different cultures from different "worlds" clash—through Western colonization or the expansion of Euro-Americans into Native American lands, for example—the physical ground is just the clay for a much larger moral struggle. It is a struggle over the very way the world out there is defined and ordered—over who rightfully rules and belongs and on what footing. The contending parties configure the "world" in such circumstances in different ways, with different conceptions of time and space. Gains and losses on the ground are of a moral as well as territorial character—the two elements are clearly intertwined.

But within the framework of multiculturalism, we have a shared world. The principles that generate conflict are, paradoxically, reciprocal ("given and received in return") and jointly held. Thus the idea that different cultures should, for example, be able to express themselves, have "territorialities of self" wherein their own histories, dress, foods, agendas, hopes, and dreams can be expressed—that is, to have some form of "self-determination" (not necessarily in a geographic sense), can be broadly agreed to. Yet sharing a principle can generate conflict as well engage different parties—as principles of national self-determination have done so for some time in, say, Northern Ireland or among Palestinians and Israelis.

In the case of multiculturalism, what we witness is the physical ground (as symbolized in monuments and parks) being opened to multiple moral universes; the land does not imply a *fixed* image of space and time, tied to a singular people. But, at the same moment, the *multi* in *multiculturalism* is not of different universes or worlds in an incomprehensible and mortal embrace but of a shared universe. A different kind of politics emerges in this context, one that mediates different moral and symbolic spaces that coexist in geographic space; but in doing so, certain "ground rules" based on "universal" liberal principles of individualism and voluntarism—a cultural identity cannot be imposed on any individual—are increasingly institutionalized. Such ground rules and principles (like self-determination) engage parties as they divide them, and divide them as they are engaged. This framework constrains as much as empowers different cultural expression and transforms cultures as much as it preserves them. (This issue is addressed at greater length in the next chapter.) In the struggle and evolution of monuments and parks, the disaggregation of physical and moral spaces and the ensuing social, political, and indeed global shifts in the contours of human association can be felt.

The Custer Monument's Last Stand

General George Armstrong Custer had the ground cut out from under him when primarily Sioux and Cheyenne warriors killed him and all his troops and Crow scouts at the Battle of Little Bighorn in Montana in 1876. And ground here that was lost was moral, as well as literal—it was a defeat, albeit temporary, of the advance and progress of America, of civilization. The ground, in the minds of Euro-Americans, was arranged in quadratic grids so it could be comprehended, harnessed, and made productive. It was a "monoculture" in the sense of defining the ground, like the Hudson River School did, as an inclusive sweeping landscape capturing the singularity of the American vision of time and place, a single story or narrative; as it looks forward, so "old" cultures were (rightfully, in that context) swept away, and absorbed, by progress. Paradoxically, American "movement" forward in time and space *had to be* rationalized, bounded, and demarcated—for in delineating "progress" it had to be measured and gauged. The United States had a dynamic sense of history, predicated on ideas of progress which were intrinsically incorporative: "history," to be understood, was conceptualized in terms of *linear time* and *linear space*—that linearity was embodied literally in the shaping and surveying of the landscape (in contrast to the cyclical, nomadic patterns of the Sioux and certain other American Indian groups). Movement through space was movement forward in history.

Under the 1868 Treaty of Fort Laramie, an extensive reservation in the Black Hills area was "granted" to the Native Americans. But by the early 1870s white miners were entering the reservation area, spurred by descriptions of gold in the Black Hills. Sioux and Cheyenne in turn, sensing a violation of the treaty, then began spending the winter in traditional hunting grounds off the reservation. In 1876 troops, including Custer, were sent to force the Sioux and Cheyenne to return to the reservation. Custer, who was just thirty-eight years of age at his death, had exhibited extraordinary bravery in the Civil War. He became, in 1863, the youngest Union general in the Civil War. At Little Bighorn, Custer's bravery proved foolhardy: instead of waiting for reinforcements, he attacked a large Indian force, apparently fearing they would otherwise escape. At the end of the battle more than two hundred troops and scouts of the Seventh Cavalry—without a single survivor—were dead.

The sociologist Joseph Rhea observes how the "Indian Wars" conjured up heroic images for Euro-Americans and were seen as synonymous with the expansion of civilization, democracy, the frontier, and the very thrust of American history itself. Rhea quotes Robert Frost's 1961 inaugural poem for John F. Kennedy: before whites moved west, the land was "still unstoried, artless, unenchanted." The first brochure of the National Park Service from the 1940s described the official and popular (among non-Indians, at least) tone through to the 1970s: "Although Custer Battlefield is a reminder of the struggle for possession of a continent, more specifically it commemorates the part of the United States Army, ever obedient to the dictates of a democratic government, played in conquering the last frontier."[12]

The American expansion westward was essentially a liberal enterprise in that it was conceived as an expansion of liberty, and the expansion of liberty was tied up with territorial expansion. The "quadratic grids" of democracy were imposed on the "unstoried, artless, unenchanted" lands of the West. The romance that occupied Euro-Americans was that of a young general who had given so much in the fight against the Confederacy—also an artless and unenchanted land, lacking in the "rational restlessness" of northern democratic ideals—and who died and spilled his blood in the ultimate civic sacrifice for his nation at the Montana frontier, in the great, open spaces and lands of the West, which themselves bespoke liberty. In Custer's mind, as well as in the minds of most of his fellow citizens, the struggle against slavery and the wars against the Indians shared a common premise about liberty—liberty is tied up in the land.[13]

Custer and his troops captured the national imagination for at least two reasons. As a military mission at the frontier they represented the thrust forward of American civilization, at the frontier of civilization both physically and morally, while embodying collective national purpose. And in defeat, the loss accentuated the conflict, the momentary containment of Manifest Destiny, and thus its heroic quality; the struggle was, in Euro-American eyes, one of Manichean dimensions between the future and the past—the future of civilization against primitive and predatory tribes. (This imagery contrasted with the "noble savage" idea of a people at one with nature, an image that was reflected in Thomas Jefferson's writings.) For the Sioux and other American Indians, it was a struggle to protect their land and their life, in every sense of the term, from marauding whites who had betrayed their promises written in treaty.

How ironic, and telling, that it was a rare *defeat* that so emotionally involved the nation and held national memory. It was an event that lived on much more so than the many Euro-American victories in the wars against different native peoples across the American continent. In national tragedy and grief, the moral mission was highlighted and collective purpose reinforced. In much the way that wrongful actions—"sinfulness"—define collective moral boundaries (and invoke righteous indignation), so defeat and loss brought into relief collective purpose and galvanized calls for communal action.

The cemetery at Little Bighorn was formally established in 1886 under an executive order signed by President Grover Cleveland, to "commemorate this engagement and perpetuate the memory of those gallant men who fought valiantly against tremendous odds." The battle represented the final military resistance, in the words of park interpretations that appeared later in the 1940s, to the "ever threatening, never-ceasing, westward march of white man's civilization." A 1956 brochure also described the battle as Indian resistance to the "westward march of civilization." However, by the late 1960s and early 1970s, the process of reinterpreting the park was under way.[14]

If a "master plan" of the National Park Service for the park in 1949 referred to Custer and his men as "symbols of . . . love of country," and their military actions as part of the "ultimate solution of the Indian problem," a master plan in the early 1970s suggested that the forthcoming centennial of the monument would be an opportune time to remind the nation that "the Indian question" was still unresolved. Instead, "the battle can be a leading banner in the appeal for re-examination of our relationship to Red America."[15] This master plan also outlined what would become, at Little Bighorn and elsewhere, an essentially multicultural approach. It would be a world where the "ground" would be layered with differently imagined moral landscapes, different tales, and different spatial and temporal mappings; all these layerings would in turn be "grounded" in reciprocal principles and rules of arbitration which allowed, in fact induced, multiple tales of identity in a shared geographic space.

The government, here in the agency of the National Park Service, takes on the role of "high-minded parliamentarian," mediating between different parties.[16] The park service's goal thus has become to tell the story of the site "objectively" and neutrally—that is, telling both sides of the story—rather than promoting heroic national and historic narratives. As such, equality between the different parties is presumed, which in turn is predicated on certain restraints of "mutual respect" and reciprocal acceptance of individual and collective self-determination. Conversely, government as the embodiment of a particular rendition of collective identity, place, and history wanes. The focus at the monument, the master plan determined, would be on the "clash of cultures," with neither culture privileged over the other. Now, in principle, many stories and many times could occupy the same space. Instead of heroic renditions of military conflict, the master plan stated that interpretations of the park should ask: Can violence be avoided when cultures come into conflict?[17]

If the park service recognized in the 1970s that the "Indian problem" was unresolved, Indians themselves (particularly urban Indians) felt that ever more acutely. (That the park service responded to issues of what the monument symbolized was itself in large part due to protests of Native Americans, such as the American Indian Movement). Manifest Destiny suggested that others would be swept into, or swept away by, America's headlong rush into history. But the Indians would not disappear, despite the terrible losses in the face of Euro-American expansion, and they were only partially assimilated into the dominant culture. Ironically, the reservations that were designed to "civilize" the Indians also proved a bulwark in preserving the continuity (albeit in a new form) of Native American life. The Red Power movement, arising in part out of the larger civil rights movement, became active in the late 1960s, with young and urban Indians (who themselves were more likely than reservation Indians to identify as pan-Indians rather than on a tribal basis) at its core. If Custer had been the symbol of Manifest Destiny, he now became for activist Native Americans the symbol of white racism and genocidal impulses. Little Bighorn thus served as a symbolic battleground, and Custer's heroic image became fiercely contested.[18]

Remembering the Indian Warriors

This struggle over the Custer heritage culminated in the changing of the name of the park, which had been a major point of contention. The previous name, the Custer Battlefield National Monument, precluded a more official, "multiperspective" interpretation of the park. An act of Congress changed the name in 1991 to Little Bighorn Battlefield National Monument. Furthermore, the new law noted that "while many members of Cheyenne, Sioux and other Indian Nations gave their lives defending their families and traditional lifestyle and livelihood, nothing stands at the battlefield to commemorate those individuals." The enactment consequently determined that a memorial would be established at the park to honor the Indian participants in the battle.[19] Although the struggle for this change had gone on for at least two decades, the moment of change was dramatic and laden with symbolism: now a people—

or peoples—were to be honored for their own culture and "traditional lifestyle" rather than representing at best a quaint and primitive society and were, in the form of the memorial for the Indian fighters, to be commemorated on an equal footing. These changes in the naming and configuration of the park, the transformation of its very symbolism, challenged and even undermined the presumption of an exclusive and inextricable tie between a singular American people and the land.

The Indian memorial did not, significantly, *replace* the Custer monument but was placed on equal footing with it, a point that was stressed through the design competition for the monument. Thus this multicultural framework is distinct from other changes in the civic arena. Changes of regime in, say, Eastern Europe bring visible changes in the way public space is organized. The overthrow of regimes often brings the physical destruction of the symbols of the *ancien régime.* A new national story or narrative is promulgated, new monuments are set up, and old ones are rehabilitated, all designed to form a new national consciousness. A new orthodoxy replaces the old.[20] In the evolutionary change towards a multicultural rendition of public life, symbols are layered on top of one another, reinterpreted, or added to, but the shift, if sudden at times, does not generally involve negation of past monuments—unless they are viewed as stained with racial exclusivity, such as the Confederate flag. (South Africa is an extraordinary case of a country where there was a revolutionary change from the politics of racial supremacy to democracy following the elections in 1994, yet the monuments and even a fragment of the anthem of the apartheid government were retained. This was done primarily at the instigation of Nelson Mandela, leader of the African National Council and first democratic president, in order to include symbolically the white minority in the new South Africa.)

The congressional mandate for the Indian memorial called for providing visitors with an "improved understanding" of the events leading up to the battle and its consequences and for the memorial to "encourage peace among people of all races." This theme of "peace" was reinforced through a subsequent decision that the design should foster "Peace through Unity," a phrase that was drawn from the words of an Oglala Sioux elder, Enos Poor Bear. What we see, then, is that the exclusive tie between a singular people and the land is broken and (as we shall see in the design of the Indian memorial, below) multiple ties to the land are drawn; for this to work in a civil society, rules are scripted to mediate between the different groups. The overarching universal values are, in this context, framed in terms of peace, coexistence, and tolerance.[21]

Notably, in the language of "peace among people of all races," a unifying and singular national theme—in spatial or temporal terms—is largely absent in the law; there is little specifically *American* (be it in inclusive or exclusive terms), beyond the narrow institutional framework, about the legal grounding and design of the new Little Bighorn monument. This is clearly a departure from past monuments. By law, six of the eleven appointed members of the memorial's advisory committee had to be representatives of Native American tribes from the area or who participated in the battle itself. Insofar as the more traditional American narrative appears—in the obelisk celebrating Custer and his men—it is now a story among stories. Indeed, Arthur Amiotte, an American Indian writer and artist who served on the jury panel for the memorial, stated that as a "tribal people of the Northern Plains," over "thousands and thousands of years and as many generations we have adapted to and learned to make peace with, live here and love this land *which at one time had no artificial boundaries.*"[22]

Of course, all humanly constructed boundaries are in some sense artificial, a figment of the imagination collectively agreed to and sacralized. But in the Little Bighorn monument and memorials we see, even on an official level, hints of desacralization of the borders of the United States to the extent that they are viewed as arbitrary and "artificial." Conversely, the boundaries of nomadic and pre-European peoples were, by implication, "natural." The process of desacralization described here is akin to Huizinga's observation, quoted at the beginning of this chapter, that every "thought seeks expression in an image, but in this image it solidifies and becomes rigid." This is true for nation-states as well, and they run the risk that by "this tendency to embodiment in visible forms all holy concepts are constantly exposed to the danger of hardening into mere externalism."[23]

The link to the land and the multicultural themes were played out in the design of the Indian memorial itself. The winning design, which was announced in 1997, consists of an earthen berm surrounding a circular plaza or "gathering place." A broad platform opens the berm to the north, and on the outer edge of this platform rest "large scale bronze ethereal tracings of Sioux, Cheyenne and Arapaho warriors," which are silhouetted against the "Great Plains sky." The two Pennsylvania designers, John R. Collins and Alison J. Towers, suggested that the memorial would connect with the existing Seventh Cavalry monument with an "unseen axis" linking the centers of the respective Indian and Seventh Cavalry monuments. In their design statement they described the memorial:

> From a distance the memorial appears to be an elemental landform, recalling the ancient earthworks found throughout the continent. An integral relationship is established with the 7th Cavalry Monument via an axis which connects the center of each element. Where this axis bisects the earthen enclosure, a weeping wound or cut exists to signify the conflict of the two worlds. Two large adorned wooden posts straddle this gap and form a "spirit gate" . . . to welcome the Cavalry dead and to symbolize the mutual understanding of the infinite all the dead possess. This gate also serves as a visible landmark and counterpoint to the 7th Cavalry obelisk . . .
>
> Selected texts, narratives, quotes, crafts, artifacts, offerings, petroglyphs and pictographs are all employed to immerse the visitor in the diverse culture of Indian men, women, and children and convey the "Peace Through Unity" message. In homage to the Plains Indians' nomadic way of life, a portion of the "living memorial" display is transient in nature and would be changed periodically to give new life to the experience.[24]

As in other battle monuments, the linkage with the earth, of body and soil, of dust and dust, is accentuated through the use of earthen berms, which symbolize Indian burial mounds. Indeed, the memorial as a whole is designed to conjure up an "elemental landform." Separately from the design statement, John Collins referred to the experience of entering the memorial as "the ground . . . rising around you as you enter," making the experience of the relationship to the earth palpable. But the symbolism is distinctly different from that of military cemeteries (including the Custer monument in its earlier rendition); there are no expressions of civic sacrifice, except in a most oblique fashion in that the memorial commemorates the Indian warriors protecting their "way of life." Nor is there any heroic description of the death of warriors immortalizing their nations, of fulfilling historical destiny. Virtuous violence is not celebrated; instead we have the "weeping wound" and the tragedy of violence. The cultural contrasts are direct—"two worlds" in conflict, the "diverse culture" of the Indians, and the homage to the nomadic way of life through transient displays, a "living memorial." The reference to nomadic ways of life and transient exhibitions implicitly expresses an opposition to the Custer monument, where history is fixed, spatially and in its destined thrust forward, its Manifest Destiny. The nomadism alludes to the "artificiality" of the borders and boundaries, internal and external, social and political, established through Euro-American settlement. Yet the cultural contrasts are explicitly balanced with multicultural expressions of "mutual understanding," with a "spirit gate" to "welcome the Cavalry dead." Collins talked of the memorial as not being "exclusive" and hoped that the two monuments' contiguity and common axis would "create a dialogue."[25]

Let's reflect on the significance of these changes: these public symbols define a new drama, a new mode of interaction. They reflect a world of multiple moral spaces, multiple times and places, yet each contained and transformed through being induced to accept the legitimacy of the other's presence. It is a "privatized public space" where the "experience" of those who make pilgrimages to the site can reflect on their own. The park service, then, seeks to appeal to and address a variety of audiences and sets itself up as a neutral protector of historical heritage. The site is also open to a variety of audiences. (The number of Native American visitors increased significantly, even dramatically, after the park's name was changed.) As such symbolic touchstones embody the "public," so public life generally takes on a more multifaceted patina.

In this picture, this portrait of life itself, history is a kaleidoscope of vistas, spatial orientations, and temporal cadences, filled with the presence of a multitude of cultures. Every culture is itself made up from a series of forms and fashions that are cultural life forms of their own. As Paul Hutton, a historian and member of the jury panel of the memorial, stated to an audience at the announcement of the winners of the competition:

> Many of the Americans who fought with Custer . . . had only been in the country for a matter of months, some of them—including the man he sent [for] reinforcements—hardly even spoke the English language; he spoke Italian. They were in conflict with people who, of course, had

> resided in this area for quite some time, for generations; and they were aided in this struggle against the Sioux, Cheyenne, the Arapaho by their allies, the Crow, the Arikara; and to the South with other columns; the Shoshone. This was a civil war not only between Euro-Americans, as they are often called now, but also between Native Americans, to decide the fate of the Western United States. It was an epic struggle, it was an epic moment, it was one of the great moments in American history; it's one of the most important battles ever fought on the North American continent; and indeed by this victory, in fact, the victors sealed their own fate because the forces of the East were then marshalled against them. Well, in our own time we have now come to memorialize the victors, and that victory, as well as memorialize those Native Americans who fought alongside Custer; the scouts who fought and died with him fighting for the future of their people.

Alluding to the desire for multiple cultures to share the experience of the park, and to the neutrality of such sites, Hutton also noted:

> And we want it to be sensitive at the same time to White America and to what they see in that place so that hopefully, hopefully in the next century—which is going to be a very different America from this century just like our century has been different from the 19th century. All Americans will come to this place which for so long has simply memorialized the cavalry and their sacrifice, and for many years, I think, memorialized Custer as an American hero. They'll come to this place and be able to see the great struggle, the heroic struggle that very diverse Americans from very different cultural backgrounds engaged in to decide the fate of a nation, and hopefully in time it, too, will be a neutral place where all Americans can come together and feel pride in the sacrifice of both sides at the Little Bighorn; that's one of the things we try to do with this Indian memorial and time will see if we have succeeded.[26]

If history is a kaleidoscope of vistas, spatial orientations, and temporal cadences, it is important to remember that even a kaleidoscope imposes pattern and order, and in that sense the past has to be reinterpreted, even (or especially) in a multicultural framework, to create some kind of reciprocity essential for a civil society. Still, the Little Bighorn site is remarkable in illustrating how different senses of time and space, how different conceptions of links to the land, shape the forms of human association, including notions of kinship, place, and belonging. Contrast the Puritan ideas of place, of a bounded territory inextricably tied to a people with a chosen destiny in a progressive and linear history—ideas that framed many nation-states as well as the incipient American nation—with Amiotte's "cyclical" description of (nomadic) Indians and the land:

> Our tribal teachings tell us we are all children of an infinite, ongoing and continual process that transcends our most basic understanding of time and place. This process is the constant interaction of the masculine celestial forces of sun, atmosphere, clouds, light, heat and winds with the potencies of an ever generous, female, maternal, fecund, green and nurturing earth and its waters.
>
> This [synergy] includes perpetual and rhythmic cycles of days, nights, moons, seasons, years and eons; all of these we recognize as only a glimpse of the great holy time of which we are a part. This process is one of fertilizing; germination; gestation; springing into blossom and fruition; and ultimately, waning and dying. Finally, we are buried or rejoined with the source of the power that created and governs these cycles. Thus it is we and all life forms are returned to the earth and transformed into remembrance. So it is as Native people in our sense of relationship that we respect and hold dear all those who have gone before us; our existing elders; our adults; our youth and those yet to be born.[27]

Fertilizing, germination, gestation, springing into blossom and fruition, and, ultimately, waning and dying: in the soil lies the cycle of death and birth.

Civic Virtue and Violence

Violence—virtuous violence—defines the outer contours of citizenship. The citizen is civilized, a civilian, a rational participant in civic life—rational in the sense that he or she shows restraint and is not prone to the passions or impulses of the moment. The frisson, the excitement (in the abstract), of virtuous violence is that it is precisely an expression of the passions; the boundaries of the restrained self are suddenly broken and merge with the civic whole, transcending daily existence and suddenly caught up in the flow of national history and destiny. But such virtuous violence is in the name of the civitas, the nation, and it is turned outward against threats to the body politic.[28] The monopolization of violence by the state—only the state, in the name of the people, can legitimately engage in violent acts—expresses the dual quality of the citizen: domestically he or she should show restraint and engage in civic (peaceful) relations with fellow citizens, while externally he or she must be ready to bear arms for the state.

This monopolization of violence by the state was central to the "civilizing process," as the late sociologist Norbert Elias described it, a process that accompanied the rise of the nation-state and, simultaneously, characterized the demise of medievalism.[29] War monuments have traditionally captured the Janus-faced quality of citizenship, celebrating civic sacrifice and the virtue of violence in the name of protecting the citizenry and its "freedom from fear." That civilizing process is now being universalized and globalized, and contemporary monuments of war and violence reflect this. In different ways, the Vietnam Veterans Memorial and the Holocaust memorial, among the most visited sites in Washington, D.C., reflect this.

Memorializing Vietnam

In 1979 a group of veterans created the Vietnam Veterans Memorial Fund to seek symbolic recognition of those who served in Southeast Asia. They hoped that by separating the veterans from the war national reconciliation could be promoted. The memorial, they stated, could make no political statement about the war. The VVMF also sought a memorial that was "reflective and contemplative" in character in a prominent and parklike setting. They also wanted the memorial to be inscribed with the names of all those killed and missing in the war. Congress authorized a site favored by the VVMF, in Constitution Gardens near the Lincoln Memorial.[30] This was a prominent site in the sacred heart of the nation, and yet this selection contained a certain unnoticed irony: the Vietnam memorial cast a different tone about civitas and citizenship than the one Lincoln had created at Gettysburg.

Constitution Gardens were conceived as a "living legacy" to the founding of the Republic and as an "oasis" in the urban landscape.[31] Nature and the Republic were once more conjoined, the Republic living eternally in nature. The park was built at the instigation of President Nixon and dedicated in 1976 as part of the bicentennial celebrations. The National Park Service hosts a naturalization ceremony each year in the gardens. Constitution Gardens thus represent an almost archetypal linkage of nature, the nation, and republicanism. Maya Lin, the designer of the memorial, sought to reinforce the "natural" setting: she chose polished black granite so as to cast back at viewers the reflections of the lawns and trees, as well as the adjacent monuments, surrounding the memorial. The Vietnam memorial thus followed, in one sense, a fairly conventional path for war memorials. But the linkage to nature was cast, and read, in an obverse form—introspective, alone, private, more akin to the poetic otherworldliness of Thoreau than the worldly civic commitment of Jefferson or Lincoln.[32] Lin wrote in her statement with her winning submission to the competition for the monument:

> Walking through this park-like area, the memorial appears as a rift in the earth—a long, polished black stone wall, emerging from and receding into the earth. Approaching the memorial, the ground slopes gently downward, and the low walls emerging on either side, growing out of the earth, extend and converge at a point below and ahead. Walking into this grassy site contained by the walls of this memorial, we can barely make out the carved names upon the memorial's walls. These names, seemingly infinite in number, convey the sense of overwhelming numbers, while unifying these individuals into a whole . . . The memorial is composed not as an unchanging monument, but as a moving composition to be understood as we move into and out of it.[33]

The reflection of the trees, the grassy surroundings, the tens of thousands of names, each telling a story, and the viewer's own visage reflected back—all conjure the personal contemplation and sadness of Mount Auburn cemetery, not the national melancholy of Gettysburg. The Vietnam memorial generates an autonomous world separate from the surroundings and draws the viewer away from the physical, civic, social, and political context and into the personal tragedies of the war. In contrast the nearby Lincoln memorial, like traditional memorials, is a fragment of the larger physical and social reality, anchoring the viewer in the civic and political life of the nation. Traditional memorials anchor the body politic, communal commitment, identity and solidarity. They are *public* memorials. The Vietnam memorial is a privatized public space. It renders visible a world of private anguish.

The sense of *private* here, in the context of personal mourning, is in terms of the root Latin meaning, namely, of the "individual alone" and "not belonging or unrelated to the state."[34] This is different from the derivative meaning of *deprivation* (or "privation"). Clearly these meanings have different political and sociological implications. In the latter case, the sense of *private* as deprived is in the spirit of the civic ideal, of republicanism as civic virtue, whereby the individual is an inherently social and political creature who is realized or fulfilled only through participation in public life. The connotation of *private* as the "individual alone" is more positive, but of greater note here is the associated meaning of not belonging or related to the state. That is to say, the individual is to be treated in and of him- or herself, not as derived from some larger body politic or community. The individual, in this setting, is a universal and not a national category. Hence the mourning represented in the Vietnam memorial is of universal import; by taking an intensely personal loss and, as such, memorializing it publicly, the pain of violence is made universal and, by extension, global. This is not the virtue of violence in the name of defending freedom and nation but violence as despair. The memorial reflects and reinforces a more contemporary sentiment to delegitimate violence in the name of the nation and to extend civic mores transnationally.

Like other war memorials the Vietnam memorial expresses the same meeting of death and life but does not, in some dialectical fashion, transcend the moment of the war itself and fold individual loss into the immortality of the nation. We as humans can imagine ourselves extending over time and space, yet here we have death, the final expression of our physical limits, our tie to the finite, to a moment, fixed, immovable, that "dark backward abysm of time." And so in death the definitive puzzle: from a dispassionate distance—a paradox. War memorials, traditionally, express *and transcend* that painful contradiction of life and death, joining the mortal remains of the soldiers in the infinite and eternal grasp of nature and nation. Instead, at the Vietnam memorial the audience is lost in the finite time of the war itself, the lives cut short, the multitude of stories, the wives left behind, the children who barely remember their fathers. Many letters left at the Wall, as the Vietnam memorial has come to be known, speak to a lost father: "I have always been told I look like you . . . how I wish I could remember you. I was four when you were killed but all I have are visions and childhood dreams of what you were."[35] All wars leave such a wake, but in the case of Vietnam national and public symbolic resolution is absent.

The deeply personal quality of the Wall is reflected not only in its design and the listing of the names of those killed in the combat but also by the public's response. Members of the public, especially but not only veterans, and family and friends of the lost soldiers, leave personal memorabilia, poems, and letters, each a story unto itself. Thomas Allen, in an introductory essay to a volume entitled *Offerings at the Wall,* writes:

> They come to the Wall silently, passing along the walkway where the black stone slabs rise. They walk slowly, seeing their faces mirrored and mingled in the rows of names . . . Freddie A. Blackburn . . . Daniel Diaz . . . Bobby Ray Jones . . . Hallie W. Smith—those 58,196 names. Usually they stop and run their fingers along the names, touching a war that this memorial keeps forever unforgotten. For those moments, they are not mere tourists . . . they are pilgrims who have journeyed to a place that has become a national shrine and an honored repository for keepsakes of grief.
>
> A few stoop and leave an offering—a note hastily written, a flag, a single rose, a burnished plaque, a teddy bear brought from a faraway time. A hat was left with a poem that said, "When we touch the Wall we know that you are there." . . . People usually leave ordinary things. The toys of sons and brothers. The badges of sons and brothers. The badges and the dog tags and medals of warriors who are parting at last with the past. Birthday cards from mothers. Notes from girlfriends growing old: Linda writes to Gary, Doug, and Billy, high school boyfriends who became names on the Wall—"After you all died . . . I pretty much screwed up for ten years . . . Now I'm much better, more responsible [but] I learned the pain and the loss never goes away. It just changes."[36]

The National Park Service, which began collecting these objects, itself noted the irony of its changing role: normally the curator selects specific objects to fit a certain narrative, in the past often in line with a national rendering of a set of events. However, as the park service notes, in this case "it is the public who collects and leaves objects at the Wall." These objects, the park service further noted, "reflect the experiences of over 25 million visitors *who each have an individual reason or story that goes with the objects* that are represented in this collection."[37] The state, rather than embodying the singular and unitary sense of the national self (as in principles of "national self-determination"), acts as a neutral facilitator collecting individual stories, while promoting the "healing" of the wounds of an apparently fractured society.

The memorial is not, then, the monument that anchors the public square, though located in the civic center of the nation. Rather, it is a venue of pilgrimage, to be visited for private contemplation and then to withdraw. The meanings associated with *pilgrimage* date from the Middle Ages: "one from foreign parts"; a wayfarer, a sojourner; one who journeys to a sacred place; "to live among strangers"; and from the perspective of Protestant Reformers, idolatry.[38] Sites of pilgrimage, such as in the most prominent cases of Rome, Mecca, or Jerusalem, do not suggest civic or national community but an almost inchoate mix of people foreign to one another. The pilgrimlike quality is also evident in the practice, which began spontaneously, of visitors leaving notes and objects at the Wall. In some respects, this parallels the practice of pilgrims lighting candles at holy sites and wedging notes into sacred structures (like the Western Wall of the Temple Mount in Jerusalem), often seeking the intercession of the Divine. In the case of the Vietnam memorial, the object and letters seek personal attachment with departed comrades, husbands, fathers, friends. Even at the Vietnam memorial religious and devotional items are among the most common objects left, together with writings and notes, rubbings of names, flags, and photographs of peoples and places in Vietnam.[39]

But the more useful parallel here is the way the idea and practice of pilgrimage organize space. The significance of a sacred site of pilgrimage is not in its territorial or geographic quality per se; it does not demarcate a territorial and sovereign jurisdiction. Gettysburg National Cemetery or the Washington Monument, for example, contains such an assumption about national territorial jurisdictions. This is why American (or British, French, and German) military cemeteries in locations outside national boundaries are designated as "extraterritorial"—such that the Normandy American Cemetery and Memorial overlooking Omaha beach, for example, is legally American territory though nominally in France. But pilgrimage sites do not make assertions about territorial sovereignty as such. Rather, they are "existential and sacred spaces," points of orientation in which territorial location (in the political sense) is, in principle, irrelevant. Such a sacred space has as a consequence a different structure than territorially bounded spaces and "admits an infinite number of breaks, and . . . is capable of an infinite number of communications."[40] The spatial imagery of pilgrimage is cyclical rather than linear, dynamic rather than stationary, and porous rather than contained.

The Rift in the Earth, the Weeping Wound

Like the Little Bighorn monument, a "rift in the earth," to use Maya Lin's description of the Wall's setting, may convey a "weeping wound." Indeed, in Lin's thinking this monument is a wound to mother nature that also evokes the healing qualities of nature. On describing going to the site of the monument before its construction she writes: "I imagined taking a knife and cutting into the earth, opening it up, an initial violence and pain that in time would heal. The grass would grow back, but the initial cut would remain a pure flat surface in the earth." Lin also talks of wanting to "cut open the earth—an initial violence that heals in time but leaves a scar."[41] This is a sensibility that is clearly different from that of traditional war monuments, like Gettysburg, where body, nation, and soil are in symbiotic and shared embrace.

Veterans offended by the absence of any reference to the sacrifice and the heroism of the troops in Vietnam referred to the memorial, with its low profile almost sunken into the earth, as the "black gash of shame." Veterans who opposed the design suggested that its black color and its sunken profile denoted shame, that it had the appearance of a tombstone, was unheroic, had no flag, and was abstract rather than representational.[42] Senior officials in the Reagan administration shared these criticisms. A compromise was reached when the sculptor Frederick Hart's lifesize sculpture of three servicemen in a more traditional representational mode were added to the memorial, together with an American flag, in 1984 (and accompanied by a vociferous exchange between Lin and Hart castigating each other's designs). The Hart sculpture, in turn, generated complaints that the women who served in Vietnam were not represented in the statue of the three servicemen (though the names of the eight women in the services killed in Vietnam are inscribed on the Wall). Consequently the Vietnam Women's Memorial was added to the Vietnam Veterans Memorial to help, it was said, the course of healing.

The criticism of the abstract character of the Wall and the absence of a representational monument (a criticism addressed through Hart's three servicemen) is telling. Abstraction, if successful in drawing attention, demands engagement and interpretation and engenders an introspective, reflective, and meditative disposition.[43] Representation contains its own narrative, and that of monuments is often heroic and patriotic—reinforced, in the case of the three servicemen, by the adjacent Stars and Stripes. Representation describes the narrative to the viewer; abstraction reflects back on the viewer, demanding that he or she take part in the storytelling. Representation describes a historical moment and draws the viewer into the larger saga of the nation (thus overcoming the fleeting, transient quality of life). It is a "re-presentation" of an event, the effect of which is to reproduce the action of the event itself rather than to show it figuratively.[44] Abstraction transcends history, seeking to capture some essential, transhistorical truth—the loss of innocence, the irreversible pain of separation from lost kin in war memorials. Georgia O'Keeffe, whose paintings were characterized by their abstract and organic themes, argued thus against the external representation of things: "Nothing is less real than realism. Details are confusing. It is only by selection, by elimination, by emphasis that we get at the real meaning of things."[45]

The Hart statue and the Vietnam Women's Memorial were, in their representative and patriotic tones, attempts to reassert a more traditional memorialization. But, paradoxically, they reinforced a more morally layered sense of America, as well as of the monument itself. If the Wall personalized violence and loss, the Hart and women's memorial introduced an explicitly multicultural rendering of the landscape. Rather than "a soldier" representing the nation, now the ethnic and gender themes were explicitly introduced. One of Hart's servicemen is black, a conscious attempt to reflect those who actually served in Southeast Asia. The women's memorial was designed to lift the "cloak of invisibility" around the women who served in Vietnam.[46] Of course, in the millions who served in Vietnam there are millions of stories, and the "cloak of invisibility" could be lifted along all kinds of categories: Protestant, Catholic, Jewish; southern and northern states; and, for that matter, southern Catholic women. And perhaps this will happen. But the larger significance is this: the monument does not, as Lincoln did at Gettysburg, serve to unify a nation, land, and polity as a single, organic entity. Rather, the land is planted with different seedlings but still grounded in common principles of equality, individual rights, and multiculturalism. One critical writer, in an article entitled "Memorial Mania," described it thus:

> What is different today is that memorials no longer represent a nation united, but one divided. They have become totems to factious causes rather than to shared ideals . . .
>
> Some, like the newly dedicated memorial to President Franklin Delano Roosevelt in Washington, stand as three-dimensional placards to political correctness. Not only is this sprawling, stone-enclosed park completely out of bounds with Roosevelt's own wishes for a memorial the size of his desk, but it has been edited to appease various special interest groups: Yielding to anti-tobacco interests, it does not depict FDR with his signature cigarette holder. In response to animal rights activists, it eliminates the fox-fur boa on the statue of First Lady Eleanor Roosevelt. And in deference to the disabled, the already sculpture-laden memorial will be expanded to include a statue of FDR in a wheelchair.[47]

In truth, multiculturalism is much more contained than this description may suggest, and it unintentionally hides the "shared ideals" implicit in this new landscape. But this description illustrates how the landscape, and its monuments, have become much more multifaceted.

At Gettysburg, Lincoln in his address, as we noted in the previous chapter, never referred to himself or the singular, nor did he refer to the dead by name. He never even referred to Gettysburg by name. His words transposed the mourning of the individual to the nation, and the sacrifice of the soldiers to the eternal life of the community. We, us, our—a shared moral landscape, a shared country. In the Vietnam memorial we have the opposite: thousands upon thousands of individual names, by the chronology of their loss in combat. The landscape, riven, has many hues.

America's Place after the Holocaust

Yevgeny Yevtushenko, the Russian poet, wrote the poem "Babi Yar" in 1961, in the post-Stalin thaw, to mourn the thirty-four thousand Jews massacred by the Nazis at that site in the Ukraine:

No gravestone stands on Babi Yar;
Only coarse earth heaped roughly on the gash.
Such dread comes over me; I feel so old,
Old as the Jews. Today I am a Jew . . .
And I too have become a soundless cry
Over the thousands that lie buried here.
I am each man slaughtered, each child shot,
None of me will forget.[48]

The worldwide memorialization of the Holocaust has universalized the horror, rather than the virtue, of violence. Civic behavior has been universalized as a norm of behavior across, as well as within, states and between communities. Certain basic rights and protections, in this context, are not predicated on membership in any territorially defined state. This, in turn, sanctions identities and communities that are "floating," so to speak, away from "home." In "Babi Yar" there is no marker, no gravestone, only the coarse earth. The earth, the land, the soil on which we stand is no longer necessarily the promise of national redemption. The African American poet, playwright, and novelist Owen Dodson wrote of the much more difficult relationship of identity and place:

I am part of this
Memorial to suffering,
Militant strength:
I am a Jew

Jew is not a race
Any longer—but a condition.
All the desert flowers have thorns;
I am bleeding in the sand . . .

There was a great scent of death
In the garden when I was born.[49]

The Holocaust is memorialized in different ways—museums, memorials, and public sculptures—throughout the world, but the United States Holocaust Memorial Museum "officially incorporated" the event "into American memory."[50] A presidential commission was set up in 1979, groundbreaking took place in 1985, and the museum itself was opened in 1993. The opening was a major event, broadcast worldwide, attended by world leaders, and accompanied by special services and commemorations in churches and synagogues across the country.[51] The site of the memorial museum itself was located at the existential center of the nation, on federal land across from the Washington Monument and facing the Jefferson Memorial.

The establishment of the Holocaust memorial museum was criticized as "political," namely, to satisfy "Jewish interests." More broadly, arguments were made that the Holocaust was not an American event, that it was a "genocidal crime committed by an alien tyranny on another continent." The memorial site, one war veteran complained, was in "the wrong place [and the] wrong country." These arguments were countered by others who suggested that the United States had a special responsibility as the "standard-bearer of freedom and human rights" and as the watchdog in a dangerous world against tyranny. Edward Linenthal, who wrote a history of the museum, also points to the dynamic history of the Mall as a "repository of American identity" and says that the National Museum of the American Indian and National African American Museum, which would follow the Holocaust memorial, will also add to the ongoing evolution of American identity.[52]

But the criticisms of the museum as commemorating an event that did not belong in America expressed a concept of place and moral space which the memorial itself was challenging (as well as what America now stood for). Similarly, the criticisms were rooted in certain presumptions about the moral associations of people and the land and the bounded nature of those associations. The debate expressed a fissure between different moral geographies. The play between people and the soil took an unprecedented form at the Holocaust memorial museum, though the changes in symbolism were only partially comprehended.

At the official groundbreaking of the Holocaust memorial earth taken from five European death camps and concentration camps—Dachau, Bergen-Belsen, Auschwitz, Theresienstadt (also known as Terezin), and Treblinka—and from the Warsaw Jewish cemetery was mixed with "native American soil" to form the museum's foundation. The redolent symbolism of the soil echoed in the remarks at the groundbreaking (the ritual of groundbreaking and the term itself are evocative of the sense that the "mediating agency of Being is the ground"). Donald Hodel, secretary of the interior, noted in his address that with "the help of colleagues . . . we personally dug the sacred soil of Terezin near Prague. It now joins those from other camps where death was their paramount duty. These holy soils are now before us to be mingled with our beloved country's rich and consoling earth. May they rest in peace in the knowledge that here, in this place today, we commence the sacred task of remembrance and teaching." The Reverend John Pawlikowski stated in his invocation, "Today we take the first step in implanting our dream of a permanent memorial into the soil and spirit of this free republic."[53]

The mingling of soils was striking in that it simultaneously preserved and transcended the symbolism of the soil for the nation. Rather than some "foreign field" forever embodying the nation in the sacrificial death of its children, soil from other lands, sanctified by the victims of genocidal violence, is brought to America. American soil enriches, it consoles, it represents hope and freedom; but in being mixed with these other soils it is figuratively universalized as well. Not only does America have a universal mission; it is itself part of and responsible to a larger human community. The soils from the camps were sacralized not through civic sacrifice, not through virtuous violence, but through *human* loss. The "lessons" to be taught concern tolerant acceptance of differences, the rights of different peoples "to be"—above all *human rights,* which by definition transcend any territorially defined state. Violence, even for the nation, is not celebrated; it is the blood and ashes of the victims which was lost, which sanctified the soil. Blood, metaphorically, casts a different meaning in contrast to military monuments. In the words of Gen. 4:10, inscribed on an inside wall of the museum, "What have you done? Hark, thy brother's blood cries out to me from the ground!"

Indeed, the groundbreaking proved to be both a forum and an impetus for the ratification of the international treaty against genocide by the United States, a ratification that had been held up for decades but was ultimately successful. The Holocaust has become a universal symbol in a number of respects: it represents a universal evil that cuts across borders and universal truths and moral values; it has come to stand for the cross-national importance of human rights and dignity; and the global role of the symbolism of the Holocaust has become significant in framing debate from Kosovo to Rawanda.

What we see is a reorganization of space, of the sacred and the profane and the national and global. The landscape, rather than a nationally bounded and sacralized "blessed plot," a primordial source of being, the organization of overarching identity, and the organic body of the nation that calls for sacrifice, is now much more heterogeneous and porous. The earth, the soil, can be sacred or profane, and national borders do not demarcate, necessarily, the sacred. The soil, the land, the country can have a much more profane (mundane) connotation: the "coarse earth" of Babi Yar and in the words of Owen Dodson: "All the desert flowers have thorns / I am bleeding in the sand . . . / There was a great scent of death / In the garden when I was born." In contrast, certain sites are global, or least transnational, in their sacredness and in their framing of social, political, and moral life. The Holocaust, in its memorials and sites, has that symbolism, and national memorials—the Holocaust Memorial Museum in Washington, D.C., explicitly so—locate the respective states in a certain (global) moral geography. (Indeed, on the day dedicating the museum, Elie Wiesel, who was central in the establishment of the memorial, turned during his public remarks to President Clinton and implored him to intervene to stop the killing in Bosnia.) The model of pilgrimage, in its organization of space sacred and profane and of sacred sites as radial "axes" (as in *axis-mundi*) unconfined by territorial, bounded sovereignties, resonates in this circumstance.

War monuments have, it was noted, traditionally captured the Janus-faced quality of citizenship, celebrating civic sacrifice and the virtue of violence in the name of protecting the people and democracy. In the Holocaust memorial we see the presentation of violence in greatly different terms, as a standard of global evil to be countered with a dedication to human rights, a kind of "global citizenship." The sense of a global morality is matched by the increasing centrality of human rights in international and national law. The spread of cross-national claims for reparations from both governments and private organizations and corporations concerning the Holocaust, but also with regard to other depredations of the past, is an expression of this new sensibility. They reflect this global morality, this sense of where civic conduct ends—or, more to the point, does not end. This is not to say that global norms did not exist before, but they were predicated on nationally defined, territorially bounded entities, nation-states, rather than individuals. Nor does this suggest that citizenship has simply been extended globally; in certain respects it has, but in others it has been utterly transformed (a matter addressed in the following chapter). We also see, in these respects, the fundamental change in the organization of violence, at least at core of global states: armies go to war not under national flags but under the collective flag of apparently *civilized* nations—the United Nations, NATO, and the like.[54]

The United States Holocaust Memorial Museum also presents and promotes, in this moral frame, a different sense of what is the "civic" in civic society. A global civic "responsibility" is reflected in part in that the museum designers sought to personalize the Holocaust. This has the effect of universalizing the experience—rather than demarcating memorials as, for example, a solely American or Jewish concern. But the designers had something else in mind with implications for civic relations: they wished to involve the visitor and for him or her to shun forever the role of bystander. This was done by graphically showing the faces in photographs of victims in the museum exhibit or giving visitors "identification cards" of individual victims as they entered the museum. The memorial was, in its mandate, to promote civic virtue and civic transformation. But the very idea of civic conduct was being transformed. It was not so much a reflection of the "general will" and of service to the nation. Domestically, the memorial rather promotes the civic virtues of acceptance, tolerance, pluralism, and compromise.[55] Civic obligation is also extended internationally, explicitly containing an expectation that the United States should act for humanitarian purposes.

One parallel between the Holocaust memorial and national monuments that preceded it is the expression of the dangers of industrialization and technology and the need to contain it—though in the case of the Holocaust memorial the dangers are of a different order entirely: the mechanization of death. In the Holocaust, bureaucracy had triumphed in its rational, scientific organization of destruction like any other modern social endeavor. It was the "expression of some of the most significant political, moral, religious and demographic tendencies of Western civilization in the twentieth century." The victims themselves were mined for resources, such as gold from their teeth. Christian and Jewish theologians, including Emile Fackenheim, Robert McAfee Brown, and Roy and Alice Eckardt, stated that the Holocaust was different from other historical tragedies in that it was bureaucratized and mechanized and thus called into question the unalloyed belief in technology and social progress.[56]

James Ingo Freed, the architect who designed the museum, brought the hard industrial quality of the camps into the design of the museum itself. Freed sought to convey the sense of alienation, disruption, forced movement, and selection in the process of human destruction in his use of raw materials—steel, brick, and glass—and in the organization of space in the memorial. Freed also designed the museum spatially and architecturally so that visitors would have to "leave" Washington, D.C. This memorial, like the Vietnam Veterans Memorial, suggested personal contemplation and pilgrimage rather than civic identification with the body politic, in the traditional sense. Freed himself suggested that it is important that "memory be sufficiently ambiguous and open-ended so that others can inhabit the space, can imbue the forms with their own memory."[57] This sentiment also reflected the museum's mandate, to "provide individuals and groups with a profoundly moving environment for contemplation and personal commemoration."[58]

Much debate ensued before and after the establishment of the Holocaust museum: Why commemorate *this* genocide, with the salience of Jewish victims (though Gypsies, homosexuals, and other groups targeted by Nazism are also commemorated in the memorial), and not others, such as the Armenian genocide of 1915?

Although Jewish communities themselves called for such a museum, this does not explain why the Holocaust resonated so broadly in the public (and why the museum is one of the most visited sites in Washington). A number of reasons can be discerned. The Holocaust as the marker of evil universalized a people's experience while simultaneously affirming "particular" identities of peoples independently of territorial sovereignty—this reflects the sanctioning of certain forms of particular identities through universal human rights.[59] (The Jews themselves embodied that paradox of a particular people expressing "universal values" inherent in the notion of a monotheistic God, a duality that the

Christians took on in a different form. In this light the Holocaust has played a special role in Jewish-Christian relations, perhaps reconciling differences through seeking forms of universalism which "pay homage," in theologians Roy and Alice Eckardt's words, to "human particularity.")[60] The Jews have been the archetypal Diaspora and as such historically marginalized. The museum in its pedagogic role has the effect of promoting minorities and diasporic forms of organization; in intent, at least, this promotes transnational forms of community as recognized and accepted forms of association across the globe. The centrality of the People of Israel, the Jews, in Western and Christian history—the Puritans were, after all, the New Israel—played a role as well in the symbolism of the Holocaust.

But it was also the industrial, technological, and bureaucratic character of the killing, in the heart of Europe, where the modern human rights movement began in the ashes of the war, which helped generate so much reflection on this particular act of genocide. The mingling of soils at the groundbreaking of the Holocaust Memorial Museum sanctified human loss. It also reflected a wider, emerging sensibility about nature and humanity's place in it. Nature, in her contours, is being internationalized.

Landscaping

In different ways and on different scales, the Holocaust memorial, Vietnam Veterans Memorial, and Little Bighorn monument are part of a "re-visioning" of the moral and physical landscape, a new sensibility of one's place here and there, locally and globally. The coarse earth, the scent of death in one's own garden—the moral uncertainty of the nation's once Edenic past—have in recent years lead to self-questioning that has threatened or unraveled the sense of a unitary (national) moral, physical, and temporal space. This is reflected in, or perhaps instigated by, revisionist histories that have questioned the unitary and essentially positive national histories among Americans, Australians, Danes, the French, and Israelis, among others, in recent decades.

In reassembling the moral landscape we, so to speak, reconfigure it physically. And so, the landscape we "need" to nurture (and which nurtures us) is changed as well. It is in this context that a "One Earth" sentiment emerges. The sense of a shared global landscape, of one planet and one Mother Earth, is reflected in the evolution of the environmentalist movement from its concern with *national* parks and preservation to *global* warming, bird migration, and biodiversity. Prominent national parks and sites are being recast in a similar way through their designation as World Heritage Sites by the United Nations, as centers of global pilgrimage.

The transformation can also be gauged in the changing orientations to travel to national parks. Consider the Grand Canyon, perhaps the crown jewel of the National Park Service. Travel to the canyon at the turn of the twentieth century was almost an expression of civic commitment. As President Theodore Roosevelt famously said in 1903, the Grand Canyon was a "great sight every American should see." Seeing the canyon was an expression of American nationalism, on the by now familiar premise that America's grandeur was incomparably better than that of Europe or anywhere else.

Comments in the 1890s in the "private visitors book" of Captain John Hance, a "guide, story-teller and path-finder" to the Grand Canyon, reveal such sentiments. One visitor wrote in 1892, "After having visited all the noted places in both Europe and America, I have seen nothing to compare with the sublimity of the Grand Cañon. I should advise all Americans to see the most splendid sight of their own country before going abroad." Wrote a visitor in 1893: "Nature's masterpiece . . . Why Americans will go to Europe and around the world, where they can see nothing to equal it, before they have looked upon this marvelous spectacle in their own land, I cannot imagine." God, nature, and nation merged: A visitor in 1895 wrote of the "sublimity of this most wonderful production of Nature's great architect," and another in the same year added more emphatically, "Doubtless, God might have made something more wonderful or more magnificent, but, doubtless he never did. America for Americans." (Following a typical remark about a Fourth of July, 1893, visit to "one of the grandest sights in the world," a different sightseer, from Philadelphia, added: "I fully agree with the above, and desire to register this statement that a pleasant lady adds much to the enjoyment of the trip.")[61]

"The Canyon became a national monument as rich with cultivated meaning," writes the historian Stephen Pyne, "as a St. Peter's Basilica or a Crystal Palace." Its geological evolution—the channeling out of the canyon by the Colorado River over millennia—illustrated how principles of progress embedded in the land corresponded with social progress and Manifest Destiny. The nationalist interpretation of the canyon remained well into the post–World War II period. Conservationists in the 1950s and 1960s sought to preserve the canyon and its environs as it represented the virgin promise of the American landscape. One prominent writer on the canyon, Joseph Wood Crutch, raised the specter that if the canyon wilderness was not preserved, the "unique privilege" of being American would be diminished.[62]

In contrast, in a National Park Service interpretive plan of the Grand Canyon in 1996, the tone had changed dramatically: "As a World Heritage Site, the Grand Canyon is recognized as place of *universal value,* containing superlative natural and cultural features that should be preserved as *part of the heritage of all people of the world.*"[63] The civic nationalism of the canyon had largely disappeared by the 1980s and 1990s. The canyon had been designated a World Heritage Site by the United Nations in 1979. By the 1980s multicultural themes had been introduced in the official interpretations of the park. The place of Native American tribes, development by whites and non-Indian groups, and the role of women in the settlement of the area were now incorporated into different programs.[64] The park represents nature globally, and the various cultural threads running through the park and the Southwest are portrayed as a series of stories rather than being encompassed in one national narrative. So, for example, Spanish exploration in the area or descriptions of the Buffalo soldiers—black American soldiers who served with the cavalry—are woven into interpretations of the area by the park service. Local Native American tribes are consulted in determining their "stories." The park service itself tries to respond to "multiple points of view" including the growing numbers of visitors from abroad—by the 1990s one-third of all visitors at the canyon are foreigners.[65] In a 1990 report, *A Diversity of Visitors,* the National Park Service states that "park visitors represent a global community" and that the park service had to find ways of providing for "people of other cultures."[66]

The canyon, rather than the embodiment of a national saga, is now celebrated for its contemplative and "intimate" qualities, demanding that the visitor make his or her own reflections and meditations. There is the curious interplay, here and in the other monuments discussed in this chapter, between universalism and the individual in his or her subjectivity. It is as if the increasing stress on "universal" or global values strips the individual of all primordial designations (except, of course, human). Even ethnicity inheres in the subjective choices people make or are said to make. External markers—color, tribal affiliation, and the like—may mark the "authenticity" of the choice, but they do not express the choice itself, which is now subjective (like determining one's ethnic affiliation on a census form). This global and subjective quality is reflected in the primary park themes of the National Park Service's "Comprehensive Interpretive Plan" for the Grand Canyon, drafted in the year 2000. It talks of the Grand Canyon's "worldwide" value "as one of Earth's most powerful and inspiring scenic landscapes, offering people enriching opportunities to explore and experience its wild beauty in both vast and intimate spaces." The canyon, home and sacred place for a number of American Indian cultures, offers us "an opportunity to consider the powerful and spiritual ties between people and place." The Grand Canyon has "sustained people materially and spiritually for thousands of years."[67]

In the erosion of the moral and sacred unity of the nation-state, different moral geographies, different "territorialities of the self," form, but they are anchored in a principle of individual "self-determination." It is a contractual arrangement, in that territories of the self depend on reciprocity—a point elaborated upon in the following chapter. And much of everyday life in such circumstances takes place in what is understood to be functional space, such as pedestrian traffic or airports—anywhere prosaic exchange takes place. In such a world, pilgrimage is the means of meeting the sacred, of seeking "eternal moments" or monuments that transcend the present, be they national parks, museums, or memorials. The parks become global representations of nature, but the experience is personalized. If nineteenth-century landscapes celebrated America, personal photography celebrates us as individuals at some sacred site—a park, a memorial—fixed in time.[68] Tourism in good part has reverted to the original meaning of the tour, "a circular movement," as in a pilgrimage to sites of transnational or global significance, rather than its more recent meaning, a series of visits for secular purposes or civic engagement.

FIVE | *Intangible Property: A Multihued Landscape*

The world has always struck me as one vast desert where my soul wanders like a lighted torch.

—MARC CHAGALL

THE "CLOSE OF THE FRONTIER," formally pronounced in the 1890s, caused deep misgivings about the future of American democracy. The rise of the city represented, in the Jeffersonian vision, a threat to civic virtue. The end of open spaces would hinder the pulse of American expansion and liberty. One hundred years later, in the 1990s, came a less noticed revolution in the contours of the landscape. Since that decade a majority of Americans live in the suburbs. In 1960 roughly a third each lived in the cities, suburbs, and rural areas respectively; at the turn of the twenty-first century, more people live in the suburbs than the cities and the rural areas combined. This form of "spatialization" of the landscape corresponds to a different political and social cadence, to different modes of belonging and place.[1]

Spatial patterns and imagery have a temporal rhythm—they organize the sense of time and movement. The city public *square* is more than a literal designation, and its metaphoric connotations go beyond that of national and civic life and politics. The public square, in fact, may be a rectangle, circular, or some irregular shape: the designation of *square* is actually more evocative in its figurative sense. The square is figuratively the nation writ small—bounded, clearly demarcated, enclosed, shared, rational, and linear in time and space. The boundaries define and channel movement, determining "insiders" and "outsiders." Pilgrimage, literally and figuratively as an organization of space, is a spatial pattern that orders the temporal rhythm in different, almost contrary ways. Perhaps it would be more accurate to say that pilgrimage is a temporal rhythm that organizes space—in a sense the cycle of pilgrimage privileges movement and temporality over space. The pilgrimage is directed towards a fixed site but assumes fluidity; it is the "point that stands still," a node, but as such it anchors a world of cyclical movement and is itself a marker of a journey through time and space of the pilgrim.

The site of the pilgrimage marks off movement; borders and boundaries are in this regard markers of movement. Contrast the borders of the nation-state: the channeling, containment, and directed flow of movement define the contours of boundaries, sociological as well as physical. In the case of pilgrimage, boundaries, sites, and markers "calibrate" the progress of the journey, literally but also in a larger religious, cultural, or some self-defined spiritual sense. A prominent illustration is the Twelve Stations of the Cross, in Jerusalem, which mark for Christian pilgrims Jesus' last journey, to his crucifixion. Historically, pilgrimage was much more of an otherworldly, even mystical exercise and consequently contained a contemplative and introspective character; in this respect as well, it is a spatial pattern that is in sharp contrast to the civic commitment of the "public square." (Thus the decline of civic commitment by Americans has been described in the context of the "naked public square.")[2] In the case of a territorial entity, such as the nation-state, the character of borders defines and determines movement; thus the regulation of immigration has been considered integral to defining the character of the nation as both a territorial and a social entity.

These relationships between bounded space and movement are clearly interrelated. Borders need movement to define themselves; movement needs markers or borders to define some kind of "progress." But there is a qualitative distinction that varies by degree: so a mileage marker on a road facilitates, promotes, or indicates movement—movement is the object of the exercise. A customs or national border post seeks to regulate, contain, or direct movement.

We see a gradual but distinct shift to the pattern of pilgrimage, where movement is gradually becoming favored. This is reflected culturally in the rise of transnational communities and diasporas that view their symbolic center—their *grounding*—as "elsewhere," in the patterns of migration and tourism, in the rise of "spiritual" and New Age movements, and the like. But this move to "movement" is not only cultural. The opening of borders to facilitate economic exchange (such as in the European Union) turns national borders into, in effect, mileage markers as the traffic speeds by. As has been commented on at length, the new information technologies also promote this phenomenon. And it is a pattern that reflects suburban life, where spatial patterns are centrifugal and centripetal, creating cyclical movements through tourism, entertainment, and other temporary forays that radiate outward from different uncontained locales, in contrast to the self-contained city. Rather than travail, travel as pilgrimage has a "rhythmic vitality, or spiritual rhythm expressed in the movement of life."[3] Even parts of cities are becoming more "suburban," and suburbs can be centers of technological and social innovation. Writes the journalist Michael Pollan: "I thought about the suburbanization of the city, manifest in freshly themed neighborhoods and malled retailing, and I thought about Silicon Valley, which in some ways represents the apotheosis of suburbia: the first time in history an important economic, technological and cultural revolution has its roots in the suburb."[4]

Suburbs have privatized public spaces, most notably shopping and recreation malls, rather than public squares. Their civic life is of a different order, more privatized with a stress on private lots and backyards facing away from the street and even gated communities. But in other ways, the suburbs have a Jeffersonian echo, with each "landholder" having his or her own property, connection to the land, and sometimes elaborate gardens of nature—which actually is not entirely an accident: Frederick Law Olmsted, among others, envisioned the suburbs as such. Suburbs also fit in a more multicultural, even global picture and one tied to the patterns of pilgrimage, transnational linkages, and migration. Suburbs are the physical mirror of the territorialities of the self, where civility is expressed in the ethics of maintaining a respectful distance and not being judgmental about neighbors beyond middle-class norms of respectability in dress, maintained yards, an egalitarian ethos, and general restraint. Homes center on the private backyard rather than the more public front porch.

Multiculturalism does not necessarily promote ethnic enclaves. Indeed, in a peri-urban and mobile world, forms of association or community are defined as voluntary, individualistic, and spatially segmented (in a social, not necessarily geographic, sense). It is striking in this context that suburbs in the United States are *less* segregated along racial or ethnic lines than urban neighborhoods. In the 1980s, as the suburbs grew, black, Hispanic, and Asian populations moved mostly into suburbs that had a white majority, rather than suburbs where the majority belonged to the respective minority. A high percentage of minorities in suburbs actually suppressed rather than encouraged the population growth of the respective minority groups in those suburbs.[5] In addition, and of equal import, immigrants from at least the 1980s—a period of high immigration—were much more inclined to settle in the suburbs than, as in the past, in urban ethnic enclaves. Nor did this necessarily indicate assimilation into some shared, homogeneous culture in the traditional sense, at least by one measure: indeed, if in the past suburbanization had previously filtered out those who did not speak English well or who had dropped their native language in daily life, this is now much less the case. The linkage between suburbanization and "linguistic assimilation" has significantly weakened.[6]

Cultural and other forms of association are segmented in that they constitute a "moral space," an "intellectual property" (metaphorically speaking) that may be experienced vicariously and virtually through different media and through different associations rather than directly. That moral space can be determined in different ways (ethnically, racially, religiously, or in gender terms), and in ways that may not be mutually exclusive. One interesting aspect of this form of community is that it is not expressed through, say, ethnic neighborhoods with their own physical, public life but rather through a private and often vicarious form of community or one experienced through tourism and pilgrimage. This is why the metaphoric use of *intellectual property* appertains to this case; the term is defined by Black's law dictionary as moveable and *intangible* property. It is not, necessarily, physical or fixed. Individuals carry their "property" with them, their own moral universes.

In early republican practice, only property holders had political rights; it was posited that property was necessary for independence and as evidence that such persons had a stake in the Republic.[7] Now there are "intellectual" property rights, where everyone has the formal right to the "protection" of his or her moral as well as material interests. It is important, and almost startling, to note the extent to which the literal and the metaphorical, and private property and the human rights of "moral and cultural interests," are interlaced. Thus the

World Intellectual Property Organization (WIPO), which is responsible for promoting the protection of intellectual property, including the mediation of disputes, and the administration of multilateral treaties on such matters, states:

> WIPO's fundamental mission is to protect intellectual property rights. It is for stimulating the creation of intellectual property such as inventions and works of art, and for ensuring the dissemination of their benefits, that the concept of intellectual property rights exists. These rights, enshrined in the 1948 United Nations' Universal Declaration on Human Rights, are worth repeating: "Everyone has the right freely to participate in the cultural life of the community, to enjoy the arts and to share in scientific advancement and its benefits," and "Everyone has the right to the protection of the moral and material interests resulting from any scientific, literary or artistic production of which he is the author."[8]

It is as if we carry our own soil, our own land, with us; we have our own personal space, a culture that is in principle voluntarily shared and experienced through different media, different associations, or pilgrimage to different countries, museums, monuments, and sites. The landscape is no longer enclosed, exclusive, and all embracing. The boundaries are more porous, and the flow of people leads to a layering of moral and cultural spaces in a common landscape. The irony is, however, that such multiple worlds generate homogeneity as well as difference, as the ground rules that support multiculturalism necessarily create, at the same time, privatized public spaces—such as malls and airports.

Privatized public spaces are neutral places that demand respect, social distance, and being "nonjudgmental." This affords multiculturalism and promotes a standardized or homogeneous environment at the same time. Some would say that such spaces are antiseptic; others would say that they are more inclusive of different forms of cultural expression and less demanding of conformity (and it would be fair to say that such spaces are both). Anyone can belong in an airport or a mall; there are distinct insiders and outsiders in the Italian, English, or Danish town square. Significantly, political activities of the civic or republican kind—protests, demonstrations, political organization, and so on—are largely absent in these circumstances. Town squares, on the other hand, have traditionally been focal points of political activism.

A Place of One's Own

The form and the temporal cadence or rhythm of privatized public spaces are writ large in the contours of society as a whole. This is apparent in the way immigrants are incorporated into the country. In the civic model, with a bounded space, a people rooted in the land—the model of the public square—the expectation has been that immigrants would progressively assimilate. The immigrant would become American in every sense of the word—politically, culturally, and territorially. The regulation of who could and who could not become a member was thus central to defining the national self, as in the principle of national self-determination. Immigration involved a bipolar shift, a shedding of the past and a grasping of an American future, not a cyclical movement (either literally or virtually) between places. Now the spatial contours and temporal rhythm of immigration, of America itself, are changing, corresponding to different ways of belonging and place and to a different kind of politics. This is reflected in the altered form and meaning of citizenship.

Citizenship, in addition to its rights and membership components, traditionally was about "belonging-in-space," as the geographer Jamie Goodwin-White referred to it, in both the literal and moral sense.[9] Similarly, rights that inhered in citizenship had connotations of place. Citizenship designates a community imagined in spatial as well as in human terms. Citizenship answered the essential questions: Who am I? To whom do I belong? What is my place, literally and morally, in the world? "I am an American" or "I am a Mexican" answered these questions. Rights, membership, space, and place are, or were, integrally related. Citizenship determined moral likeness and proximity and moral distance. Citizenship was, in principle, bounded and exclusive, determining who belongs and who does not; determining that boundary has been a dynamic process. Part of that determination is through the selective acceptance and rejection of outsiders into the community and polity. Once incorporated, the outsider can be naturalized, made like us, to share our space, communally and morally. In sum, citizenship describes in good part a *spatial process* of delineating boundaries and borders, insiders and outsiders.[10]

It is in this context, we can also begin to understand the significance of the much more recent, contemporary transformation of citizenship which has undermined the classical sociological underpinnings of citizenship. The traditional role of citizenship in directing loyalties, in determining the "one compulsory community" that is the nation-state, in promoting civic and republican participation and in designating moral as well as physical "place" is being been undone. The problem with a discrete state, and the multitude of associations that arose in tandem with the republican state, was that the state or government was not the only agency in the community—and herein lies the significance of traditional citizenship for sovereignty. Sovereignty, as we noted earlier, ensures that a territorially defined political community can exist at all. Citizenship expressed ultimate loyalties and overarching identities. Constitutional rights, and the jurisdiction of the courts, "end at the water's edge," at the national boundaries. The state, classically, was attributed with legal personality (practicing "self"-determination, or "self"-defense), with policy on foreign relations and cross-border activities (such as immigration) a prerogative of those agencies that represented "the people"—the legislative and executive branches of government.

Contemplate the "typical" nation-state, at least until recently. Its members are citizens twenty-fours hours a day. They have to get permission to leave and enter (through passports). The citizen needs to be registered at birth and death, as well as various events in between, such as marriage. He or she has to contribute to some kind of pension scheme, pay taxes, obey the law on a round-the-clock basis, and even give his or her life if necessary. In most nation-states, until recent years, the citizen had to abjure loyalties to any other state. The state is sovereign and, in principle, the source of primordial identity: one is an American or a German *all the time.* Except in rare circumstances, one was compelled to belong if designated a citizen through birth or kinship. The nation, embodied in the state, was the "consistent self-conception" of the individual. The temporal and spatial commitment was almost total. The extent and variety of flows that were regulated by the state were—again, at least until recently—very broad. The state controlled the flow of people, goods, and even information and tried its utmost to keep violence outside its borders. Only a designated few could claim to represent the nation domestically and, especially, externally. The very distinction between "domestic" and "foreign" or "international" reinforced the organizational boundaries of the state.

Organizations and different social categories ("bounded categories") filter the nature of interchange: who gets resources, who gets to be included as "one of us," who represents and who follows, who may interact and on what issues, even who one may marry, and so on.[11] The state, above all, determines citizenship, which channels access not only to resources (such as work and passports) but also claims to "identity" (as an American, Dane, or Botswanan). Although a bounded category like citizenship is of an abstract character, it is ultimately expressed on the interpersonal level—even if it is as tourist, illegal alien, border control officer, worker, or employer. Territorial control ensures the near totality of temporal and spatial commitment. The emotional commitment and the sense of affiliation is (or was) extensive indeed.

With the growing prominence of transnational ties, human rights, and dual citizenship we are seeing the state as less the "one compulsory community." The legal demise of citizenship as a singular loyalty and the constraints on states with regard to citizenship and human rights have significantly eroded the representation of the state as the "generalized self" and as the embodiment of a unitary nation. Consequently the representation of states as legal personalities with unitary national identities is in question.[12] Globally, this undermines the national self-other distinction (as in national self-determination)—that is, the premise that one's identity *has to be* fixed in one national and territorial jurisdiction, based on a notion of mutually exclusive "insiders" and "outsiders"—and creates the basis for transnational associations and "fields" of social relations. That is not to say that borders as filtering mechanisms do not remain very important but that social and legal fields are created for cross-border activities—from areas in commerce, ethnic association, the environment, gender, religion, to sports—which are, so to speak, autonomous of the structuring effects of states. It is in this context that we can carry our "intellectual property," our own moral properties or universes, with us. It is a form of spatial organization which also encompasses a more dynamic tempo, a more mobile and fluid trajectory to peoples' lives.

Once it was suggested that dual citizenship was akin to polygamy-one could no more have loyalties to more than one country than one could have loyalties to more than one wife. This is no longer the case.[13] As one Mexican American commented to a newspaper reporter outside the Mexican consulate in Houston in 1998, after regaining her Mexican nationality under a newly legislated Mexican law allowing dual nationality: "I love both countries . . . [Before dual nationality] it was like I was being asked to choose between my mother and my father."[14] Transnational and diasporic communal identities and affiliations are clearly much more in evidence, facilitated by the ease of global communications, mediating new forms of virtual and vicarious communities.[15] Such developments could happen only with the unraveling of the "inextricable" and exclusive tie of "the people" to the land; citizenship no longer designates the seamless conjoining of territory, people, and polity.

The social landscape has, then, become more complex. Bordering has become much more multifaceted, takes on both geographic and nongeographic forms, and may be of a social, political, or economic character. Political institutions (such as human rights), ethnic or other forms of community, and territorial states are no longer necessarily coextensive. Borders, in the larger sociological sense, also designate inclusion and exclusion and, particularly with ethnic, religious, and other such communities, moral proximity and moral distance. It is at the interplay of these spaces or territories—moral, social, physical, economic, and so on—that we can discover incipient political forms. Administrative and judicial rules are becoming ever more significant in mediating and regulating the kaleidoscope-like complexity of the world. The logic of rights is increasingly constraining the logic of popular sovereignty, civic republicanism, and, by extension, the legislature.[16]

Whereas in the past incorporation was conceptually a relatively straightforward proposition of assimilation and acculturation into an imputedly common, shared, and homogeneous space, a "community of character," with a shared collective self and sense of place, now the issue is, in theory, more complicated. With multiple spaces and multiple points of departure, the physical or geographic domain we share is open to multiple claims. Clearly the model of assimilation becomes, at best, problematic and, at worst, demonized. Similarly, the bimodal, unilinear notions—in which the "host" society "absorbs" the "immigrant"—become a problematic vocabulary; what we see instead is a transnational and more dynamic idiom, in which "social spaces" are negotiated. The domestic arena of states is visualized as much more polymorphous and porous than homogeneous and bounded. Citizenship and, more generally, rights do not raise to the same extent as was once was the case any expectations of "belonging" in any exclusive sense. This description raises the question, then, of what the ground rules, or organizing principles, are for such a framework both on the level of social exchange and on the level of its political organization.

A frame based on reciprocal contractual agreement, global in scope but worked out (albeit unevenly) through national and regional institutions, is emerging—but not in the traditional sense of the economists. That is, we do not witness self-interested, self-defined, and independent actors (be they individuals or organizations) rationally constructing mutual agreements in a vacuum but a social and political framework of immense coercive power mediating between different spaces *using the logic of contractual law.* It is a contractual logic anchored in certain moral assumptions of the dignity and the privileging of the individual and private property over all other ideological arrangements; this logic assumes that collective identities are legitimate only insofar as they reflect certain voluntarist understandings, and it reinforces a tendency toward "privatization," of property and of identities. It is in this context that one can get a nuanced sense of "privatized public spaces" in its moral or physical forms. It is in the mall, the airport, or, more broadly, suburbia—and "surburbanized cities"—where individual "territories of self" are mediated.

In this new framework individuals or groups can "negotiate" social and moral spaces based on their "rights." Human rights, and rights generally, in this sense are enabling, or empowering. This framework is transnational in that "identity claims" do not have to be situated in the claim of shared national identity (as in the case of more traditional citizenship practices, in which the struggle was for *inclusion* and integration—as in Martin Luther King Jr.'s call to judge people on the "content of their character," not their skin color). Diasporic claims of being, even if mythically, from "somewhere else," with one's own history, agenda, and grievances, are an increasingly legitimate form of expression. It is striking how, while the state is "denationalized," claims to certain forms of identity, usually but not exclusively of an ethnic character, become ever more salient, but such identity claims are by and large not predicated on territorial separation. Although state and "nation" (defined ethnically or by diasporic homelands, for example) are bifurcated, they respectively remain and perhaps grow in significance, but they constitute each other in different ways than in the traditional nation-state.

Yet there are distinct ground rules that are of an immensely coercive and constraining character. Human rights and rights generally become a critical mechanism in mediating the multiplicity of "social spaces" no longer necessarily delimited by geographic demarcations. An ethnic group or its imputed representatives can make claims on government in terms of its human rights—regarding the content of education or freedom of expression, for example—and in this sense human rights are enabling. In doing so, a set of contractual demands are being made—here, the social contract of human rights which privileges individuals and voluntary collective identities. Through such claims, "we" (be it an individual or a social group) bind ourselves not only in the nominal sense of a contract but also structurally into a set of institutional scripts and rules. Even if the initial action is voluntary (for example, claiming human rights), "our voluntary cooperation creates for us duties [that we may] not have [initially] desired." Thus the "law confers rights and imposes duties upon us as if they derived from a certain act of our will."[17]

For all the talk of multiculturalism, the degree of diversity which will be tolerated in the name of human rights is, in certain respects, remarkably constrained—constrained by the *voluntary* and *individualist* presumptions inherent in human rights. Thus the threat of female circumcision is a legitimate basis for claiming asylum; similarly, arranged marriages cannot be enforced—and the courts have ruled on this—on unwilling parties, whatever the traditional culture may desire. Human rights claims change the nature of identities, and contracting into human rights when it is at a party's advantage has certain possibly undesirable outcomes.[18] Tradition—the particularistic identities nurtured by multiculturalism—is not so much "indigenous culture" from time immemorial but "invented tradition" or "edited tradition," edited to suit the confines and moral presuppositions of human rights.[19]

Thus the courts and administrative bodies come to play an increasingly prominent role.[20] The judiciary as an institution is especially well placed in the context of rights and multiculturalism in that it is the forum through which rights claims are adjudicated. Furthermore, boundary drawing and demarcating social spaces are central to the very raison d'être of the judiciary: it places bounds on the actions of government vis-à-vis the individual, and it delineates personal distance between the state and the individual and between individuals themselves. Conversely, legislative politics, in the republican, civic model, has been about the collective interest, in which commonalities (or at least majorities) are stressed. In stressing the "collective interest," the notion of personal space is delimited. Individual political rights in this context are designed to promote participation, conjoin the voting populace in a shared public and national identity, and define communal and national goals.

The increasing importance of the judiciary since the 1970s is illustrated in the following stark figures: the caseloads (commenced cases or cases on docket) of the federal courts, involving both criminal and civil cases, jumped dramatically between 1970 and 1995. This leap in the caseload is true for all three levels of the federal courts: in the Supreme Court the caseload went from more than 4,000 cases in 1970 to more than 7,500 cases in 1995 and in the court of appeals from more than 11,500 in 1970 to almost 50,000 in 1995. In the district courts in the same period, the caseload went from about 125,000 cases in 1970 to approaching 300,000 in 1995. The number of cases granted review or acted upon also, generally, reflect an upward trajectory. These figures, if anything, hide the extent of the growing role of the judiciary and, furthermore, do not reveal the degree to which the expanding web of rights drives this process. For example, the more recent appearance of a multitude of arbitration bodies of different kinds in both private and public organizations, discussed below, and the creation of special and quasi-judicial courts (which concern areas from domestic violence to homeowners) are not included in these figures. The caseloads of state courts also do not appear in these numbers.[21]

The growing role of rights is evident in the following figures: from 1945 to 1960 the term *human rights* appears in only 68 cases and the term *civil rights* in more than 1,000 cases; from 1961 to 1970 the respective figures are 159 cases and more than 3,500 cases. Then we witness a surge: from 1971 to 1980, the numbers are 861 and more than 15,500 cases respectively; from 1981 to 1990, 2,224 cases make reference to human rights and more than 33,000 to civil rights; and from 1991 to 2,000, human rights are noted in some 7,000 cases, and well over 45,500 cases make reference to civil rights.[22]

Many administrative institutions and mechanisms come into play in this period as well. Government is increasingly involved in the regulation of the workplace and of both private and public organizations, most explicitly in creating a more "inclusive," nondiscriminatory, and multicultural environment along ethnic and gender lines. The shift in government regulation of the workplace is revealed in the following figures: between 1900 and 1964 one administrative agency was created to regulate businesses; ten agencies were created between 1964 and 1977.[23]

But this process of the growing judicial and administrative role and character of the state cannot be reduced to court cases or litigation. National and international laws and understandings are cascading downward, leading to and reinforcing a surge of rule making and regulations that are often responding to larger national, regional, and international judicial and administrative exigencies, not only in government agencies but also in private and corporate institutions.[24] Thus, this process goes beyond high-profile court cases: it is explicit in public and private organizations' rules on "diversity," "hostile work environments" (regarding both ethnic and sexual matters), and sexual harassment. It is in this larger sense that we must understand the "judicialization" and the massive growth in administrative rules and the overarching role of human rights, rather than narrowly focusing on actual litigation and the actual litigants (though that is, of course, centrally important). Although less noticeable, hearings within private and public organizations have surged as well, alongside and in response to growing judicial review (this is true in other parts of the world, including Europe). This process is paralleled by the near end of presumptions of assimilation that we "should" all melt into a common, homogeneous culture.

The judiciary and administrative bodies also serve as "traffic cops," trying to adjudicate a social world of potentially conflicting spaces. The issue of adjudication becomes all the more critical as spaces, no longer geographic as such with fixed demarcations, constantly have to be negotiated. The judiciary and administrative rulings reinforce, or create, a set of rituals that "respect" social spaces and thus create or maintain social order. The everyday rules that govern "street traffic"—neutral and demanding no sense of felt commonality—are, when taken to a higher institutional level, a way of adjudicating a society in which the presumption of shared and implicit norms and understandings (beyond keeping one's distance) is largely absent. It is akin to pedestrian traffic on the sidewalk or in the mall, where intricate procedures are maintained to avoid bumping into one another or, more broadly, to avoid any engagement beyond the momentary and superficial.[25] Although such exchange of avoidance appears to be shallow, it also allows individual parties to maintain their privacy, their "territorialities of the self," and their own worlds without being scrutinized, at least overtly. Such a framework also promotes a modicum of civility (in the sense of the absence of violence) through disengaging potentially conflicting "territories."

When personal and group space is so conspicuous, and when personal markers, the "territorialities of the self," are displayed through dress, speech, jewelry, ethnic insignias, accents, and body language, the rules of public engagement also need to be more elaborate, explicit, formal, and distant. Politeness and etiquette are used to mark (civil) distance, not closeness, and are critical when social spaces are constantly in contact, fluid, portable, yet defined. "Street traffic," neutral and predicated on avoidance rather than engagement, is the metaphor, and the actual reality at times, of civic life in this context.[26] Rules become all the more elaborate because in these fluid circumstances cultural borders are constantly coming into contact, unlike geographic borders, which are fixed. Personal and group spaces are carefully guarded, legally as well as socially, against affronts, and "self-determination" takes on a whole new meaning in comparison with its traditional national usage: more personalized, less abstract on the everyday level.

Rules of avoidance are also illustrated in the reverse: different cultures should be respected and, when discussed overtly, celebrated. Conversely, critical comment is seen as an affront. In a multicultural frame, public and civil relations are "acknowledged through avoidance . . . rituals—that is, through expressions of respectful ceremonial distance, both physical and symbolic."[27] This is the reverse of the "public square" and of the classic civil society model.[28] Ethnic, social, and religious community is segmented and privatized and, as noted, often vicarious and mediated through different means. In the case of ethnic communities, the irony is that different forms of ethnic community as such are less likely to be geographically contiguous than in the past, especially with the growth of a suburban society.

Painting the Colors of the Landscape

Citizenship customarily was in part a spatial process of delineating borders and boundaries, insiders and outsiders. Traditionally, borders and boundaries were understood to be coextensive, which is to say that the regulation of the flow of people and the definition of communal boundaries were territorially constituted. Naturalization and citizenship designated a shared space and place—the soil and the people were, so to speak, one. As a consequence, the question of immigration implicated issues of assimilation or the absence thereof. What is striking is that in debates on immigration in the United States between restrictionists and liberals, from early in the twentieth century through at least the Hart-Cellar Act of 1965 (which ended ethnic quotas in immigration law), there was a shared assumption of the importance of citizenship designating a unitary nation. The restrictionists objected to groups they viewed as "unassimilable" and hence which should be excluded, while liberals believed such groups could indeed be assimilated into a common American identity. Similarly, citizenship implied a singular loyalty until roughly the 1980s, when the U.S. government began to tolerate more easily multiple citizenship.[29] A similar stress on assimilation has been, and in parts still is, evident in Western Europe, insofar as states there consider themselves "immigration countries" at all, such as France.[30]

When we examine how the courts have actually "constructed" social and physical boundaries, we find remarkable changes over time in notions of citizenship, space, identity, and place.

When, for example, one considers U.S federal court cases on immigration and citizenship from the 1920s though World War II and beyond, the social, cultural, and communal boundaries are understood implicitly and explicitly to be defined by the national borders. Assimilation and exclusion are presumed to be essential to an orderly society. "Differences" to be "negotiated" are in principle of an international, not a domestic, nature. And defining *race* and *ethnicity*—which are frequently understood to be cultural rather than genetic concepts by the courts in this period—is critical in determining who can join the territorially defined body politic.[31] This is in contrast to the courts' more recent role in dealing with race and ethnicity issues—now the courts find themselves making delineations as to internal or domestic cultural demarcations.

For instance, in a case that took place in 1944, *Ex Parte Mohriez,* the court sought to define *whiteness* in order to ascertain whether a man named Mohamed Mohriez, born in Sanhy, Badan, Arabia (today Saudi Arabia), and termed an Arabian, was eligible for naturalization. This was one case in a series in which plaintiffs sought to be defined as white—in addition to being considered Hindus, Moslems, Arabs, and so on—in order to be eligible for naturalization. Congress had, in its first formal pronouncement on citizenship, restricted naturalization in 1790 to "white persons." This racial requirement for citizenship remained in force until 1952, though other requirements for naturalization had changed frequently. As immigration levels reached unprecedented highs at the turn of the twentieth century, people began to argue about their racial identity in order to acquire citizenship. Applicants from China, Japan, Burma, and the Philippines and those of mixed-race background failed in their efforts. Mexicans and Armenians were classified as white. Then there were those whose "whiteness" was on the liminal margins, as far as the courts were concerned, and in such cases the courts wavered.[32]

What is interesting about such cases is how the issues of "whiteness" were a function of cultural concerns about assimilation, not genetic issues per se. This could help, or hurt, the chances of the plaintiff. The judge's reasoning in the *Mohriez* case is enlightening in this regard:

> The question whether one is a "white person," within naturalization law, is a question to be settled in accordance with the understanding of the common man and turns on whether [the] petitioner is a member of one of the races inhabiting Europe or living along the shores of the Mediterranean, or perhaps is a member of a race of Asiatics, whose long contiguity to European nations and assimilation with their culture has caused them to be thought of as of the same general characteristics . . .
>
> Mohriez was on January 15, 1921, admitted to the United States as an immigrant coming here for permanent residence. Mohriez regards himself and his immediate ancestors as being white persons belonging to the white race. He himself spoke Arabian as his native language . . .
>
> The question whether one is a "free white person" within the naturalization law has now become a question to be settled "in accordance with the understanding of the common man" and turns on whether the petitioner is a member of one of the "races (a) inhabiting Europe or (b) living along the shores of the Mediterranean" or (c), perhaps, is a member of a race of "Asiatics whose long contiguity to European nations and assimilation with their culture has caused them to be thought of as of the same general characteristics."

The decision goes on to make a comparison between Arabs and Jews, which has certain poignancy in the light of subsequent history in the Middle East, and draws the Arabs into the sweep of European history and technological progress:

> In the understanding of the common man the Arab people belong to that division of the white race speaking the Semitic languages . . . Both the learned and the unlearned would compare the Arabs with the Jews towards whose naturalization every American Congress since the first has been avowedly sympathetic . . .
>
> As every schoolboy knows, the Arabs have at various times inhabited parts of Europe, lived along the Mediterranean, been contiguous to European nations and been assimilated culturally and otherwise, by them. From the Battle of Tours to the capitulation of Granada, history records the wars waged in Europe by the Arabs. The names of Avicenna and Averroës, the sciences of algebra and medicine, the population and the architecture of Spain and of Sicily, the very words of the English language, remind us as they would have reminded the Founding Fathers of the action and interaction of Arabic and non-Arabic elements of our culture. Indeed, to earlier centuries as to the twentieth century, the Arab people stand as one of the chief channels by which the traditions of white Europe, especially the ancient Greek traditions, have been carried into the present.

It is interesting that Judge Wyzanski, who wrote the decision for this case, had a clearly liberal view of immigrants and immigration yet stressed the assimilatory aspects of it. A unitary and culturally homogeneous nation did not necessarily demand a restrictive policy. And so he ends his decision with a ringing call:

> The general policy has been to say that if a person is admissible to the United States for permanent admission, he is susceptible of naturalization. This policy has obvious advantages. A person ought not to be encouraged to come here to live and to have children born here, who under the Fourteenth Amendment will automatically become citizens, unless we are prepared to give him the advantages of, and expect him to assume the obligations of, United States citizenship. It is contrary to our American creed to create a superior and inferior brand of permanent residents.
>
> And finally it may not be out of place to say that, as is shown by our recent changes in the laws respecting persons of Chinese nationality and of the yellow race, we as a country have learned that policies of rigid ex-

> clusion are not only false to our professions of democratic liberalism but repugnant to our vital interests as a world power. In so far as the Nationality Act of 1940 is still open to interpretation, it is highly desirable that it should be interpreted so as to promote friendlier relations between the United States and other nations and so as to fulfill the promise that we shall treat all men as created equal.

The judge accordingly granted the petition for citizenship.[33]

Other cases, in which claims to "whiteness" were rejected, similarly based their decision on issues of assimilation (and, in such cases, the apparent likelihood that assimilation would *not* occur).[34] Thus in both circumstances, whether the response was positive or negative, the concern was about belonging and about creating a common, shared social and cultural landscape within the physical borders of the United States.

Other cases concerning ideological loyalties, not ethnicity, also presumed and stated the same premise, that of assimilation into a common American *political* culture.[35]

However, from roughly the 1970s and 1980s we can discern a distinct change. In addition to the upsurge of legal references to human and civil rights, there are the novel references to terms such as *multiculturalism* and *diversity*. Two developments are of interest: first, the continuing interest in cultural, ethnic, and religious practices but with a twist—cultures are now viewed not in terms of their likelihood of assimilation into American society or lack thereof. Rather, the question is how, and to what extent, claims of different cultural practices can be accommodated. In extreme cases, the culture defense has led to the acquittal of what would otherwise be considered heinous crimes by the courts, such as the case of the Japanese woman who drowned her children in the sea off Santa Monica. Saved from apparently also trying to drown herself, she was given a reduced sentence for manslaughter (five years probation), as it was argued that a mother-child suicide pact was an honorable response in Japan to a husband's infidelity. In contrast, Susan Smith, the South Carolina woman who drowned *her* two children in a lake, received life in prison.[36] Courts and lawyers go so far as to make anthropological inquiries in order to ascertain the authenticity of cultural claims (such as in the Nebraska case discussed below).

However, the more prominent cases capture the constraining as well as empowering dimensions of human rights. In this sense, we see what we see with most social and institutional constructions: the rules that are essential for social *engagement*—a common idiom, vocabulary, or syntax, so to speak—are simultaneously *constraining*. Curiously, in a dialectical fashion, avoidance is necessary for engagement in a multicultural system (rules engage as they divide, and divide as they engage), or to put it in plainer language, to have difference, we need a system for *mutual* recognition of difference. If I am to recognize your "differences," you have to recognize mine. This, almost by definition, demands individualist notions of rights and of community. Culture is edited in this context, according to the voluntarist presumptions of human rights noted earlier.

Two cases that received much publicity are notable in this regard. The Nebraska case involved an Iraqi refugee from the Persian Gulf War who arrived in the United States in 1995 and was convicted of first-degree assault on a child. The refugee, Latif Al-Hussaini, thirty-four years old, "married" a thirteen-year-old girl on November 9, 1996, who had been "given away" by her father. After the ceremony, Al-Hussaini took his "bride" to a new home, where he engaged in sexual intercourse with her, despite her objections. Al-Hussaini was later arrested but claimed he did nothing wrong because arranged marriages with young girls were customary in his country and legal under Islamic law. The court sentenced him to four to six years' imprisonment in September 1997. In the court's decision we see the constraining aspects of human rights, namely, that any cultural practices are legitimate but only to the extent that they do not violate individual rights. Thus, as noted earlier, in the very act of making human rights claims (including, for example, asylum status), as immigrants and ethnic or religious groups and individuals repeatedly do, so they obligate themselves, unwittingly perhaps, to a web of legal rules that may not always serve their purposes.

The Nebraska case is telling in another respect as well. It illustrates how the legal system edits or reinvents culture not only coercively but also through redefining what that culture (or, in this case, religious culture) is in its purportedly true, authentic form. In Al-Hussaini's appeal in 1998 against the state of Nebraska, the prosecutor for Nebraska argued that, to quote from the judge's decision, "while Al-Hussaini attempts to lessen his culpability by claiming that the acts were sanctioned by religion, there is evidence in the pre-sentence investigation report that such a marriage would not be universally accepted in

Iraq and, in fact, was highly unusual and would be considered wrong in the majority of Iraqi communities."[37] The court of appeals affirmed the sentence of the lower court, noting that "there is really only one victim of this crime and that is the 13-year-old child with whom Al-Hussaini has sexual intercourse without her consent." This case is an interesting contrast from the past, in that the court takes on a global perspective, placing itself in the role of an imaginary anthropologist investigating different world cultures. That is an innovation from the traditional judicial position of not crossing national borders; as in the *Mohriez* case courts concerned themselves with the domestic compatibility of a "external" culture, not the culture's "authenticity" in some foreign, indigenous homeland.

Another case, heard before the Board of Immigration Appeals in 1996, determined that female genital mutilation of a coercive nature was grounds for granting asylum. The court noted in its decision the coercive nature of the marriage (into a polygamous union) of the applicant, Fauziya Kasinga, and how she was pressured to undergo female genital mutilation. Here the court makes no bones about its abhorrence of a certain cultural practice, in this case of the Tchamba-Kunsuntu tribe in Togo. Mutilation of this kind, in which in its extreme forms the "female genitalia are cut away" and the "vagina is sutured partially closed" and which causes serious, sometimes life-threatening complications, is persecution under the law. The Immigration and Naturalization Service, fearful of an avalanche of asylum claims on this basis, asked to limit such claims when tribal members believed "they were simply performing an important cultural rite that bonds the individual to the society" and to exclude past victims of female genital mutilation if they "at least acquiesced " to the rite. Again, the "voluntary" quality of the act, even for the INS, becomes the touchstone.[38]

The courts demarcate the boundaries (in a nonterritorial sense) of groups not only by editing cultures but also by defending them from assault—sometimes literally. So it is no surprise that in this period we see the growth of hate crime laws. In the United States, all but nine states now have hate crime laws, which ratchet up the penalties for crimes motivated by prejudice against ethnic, religious, and other groups. The federal government also has a variety of laws and offices now in place to protect ethnic, religious, and gender groups.[39]

We are not witnessing the unraveling of community so much as its reconstitution in new, often nongeographic, polymorphous terms. The state still shapes identities, by establishing boundaries and ground rules of acceptable activities.[40] That identity, however, is not constituted territorially, in that a people and land are viewed as exclusively and inextricably linked. And this has changed patterns of incorporation, to one more "interactive"—that is, groups (invented, imagined, or historical) can "negotiate" a social identity, recognition, and place that are not predicated on assimilation. Similarly, diasporic identities are no longer on the margins—legally or socially—as they once were. And yet there are also obligations—to pursue "one's own ends" in certain constricted paths; by the very terms by which groups can claim autonomy they restrict themselves to essentially individualist and voluntarist premises about identity and community.

In this way the state, though in gradual, piecemeal fashion stripped of its claim to embody a singular nationhood, still plays a crucial role in forming the boundaries of identity. Thus multiculturalism turns tradition on its head: it is "progressive" to espouse tradition, but that tradition should be progressive, that is, informed by presumptions of human rights. Tradition is "edited" and "reinvented" to reflect the contemporary world. To paraphrase the historian Joseph Levenson, tradition changes in its persistence as well as in its rejection.[41]

Virtual Citizenship

It has been observed that where organizational control is relatively weak, cross-boundary ties may be more multifaceted—the organization may display less solidarity, and less emotional commitment, but may also transmit more information from different sources more easily.[42] The organizational control of government is still considerable, but, given developments in recent years and decades, the ability of government to control cross-border flows of different kinds and consequently maintain "emotional commitment" has weakened. Furthermore, the relative importance of different state agencies has shifted around—as we observed, administrative and judicial agencies play a more significant role, and more broadly judicial rules suffuse daily life from work to family in a way that would have been foreign to citizens in early- and even mid-twentieth-century America (or Europe for that matter). Where does this leave the individual vis-à-vis "civic society"?

Georg Simmel, the German sociologist referred to in the introduction, observed that smaller groups will have qualities, including certain forms of interaction, which would disappear in larger groups. He recognized that the larger groups were, the more likely that formal rule—law—and agencies of different kinds, from the motor vehicles bureau to the court system, would grow in significance. Simmel also noted that with increased social complexity or differentiation in the division of labor of society, the individual would have ties to a growing number of social circles that intersect but would be tangential to one another. Thus, as individual commitments become extended across a number of social circles, the scope of individual "freedom" is increased, social relations generally become more impersonal, and the consciousness of a discrete "self" is heightened. The number of groups the individual can join is larger. Membership is based on rational (interest-based) criteria and is largely voluntary. Even marriage becomes one commitment tangentially intersecting with many other commitments, likely contributing to the increased divorce rate. Social and geographic mobility increases. The city, representative of this phenomenon, portrays the paradox of increased, yet more impersonal, interaction. Simmel, talking of early-twentieth-century industrialization, saw the life of the individual becoming more segmented and social relations becoming more fragmented.[43]

With globalization it would appear that this phenomenon is expanding further. The state is less the "one compulsory community" as in both national and international law dual citizenship and multiple allegiances are sanctioned. And we see on the global stage an increasing formalization of transactions that are less likely to be through official channels of the state: international law, including international private law, is becoming much more dense.[44] International nongovernmental organizations are much more common, indicators of a more diffuse environment in which state boundaries are less intensive (even if, at the same time, states serve as nodes for the expansion of global transactions). Individuals can, and do, have "social circles" that circumvent the state—from sports and entertainment to high culture, from work to (most dramatically) different citizenships. In this vista, it would be hard to imagine countries, including the United States, maintaining the same kind of emotional commitments they have had in the past. Even the military sociology changes: national service and readiness to die for one's country fall away to a volunteer force, a shift that is even seeping into European countries.

It is in this context that one must view the current discussion among social scientists and other social observers about the "decline of civic society." The argument most prominent in the public eye is that of Robert Putnam in *Bowling Alone.* Putnam and others argue that civic society is in decline as a result of decreasing participation in civic groups and volunteer activities from parent-teacher associations to the Boy Scouts, the drop in voting in elections since the 1960s, and other measures of apparently weakening social trust.[45] Putnam's measures on civic participation are contested—some suggest that participation in voluntary groups may have shifted to forums that are not formally counted in some way (informal book clubs, for example) and which belie any "decline."

What is interesting is that the measures that are not in dispute—voting, for example—*do* suggest a decline in republican and civic politics. People are not engaged in the matters of the Republic, in the primary senses we can measure, to the extent they once were. But what is interesting about this does not stop there: the drop in voting and traditional politics (for example, membership in trade unions has decreased, identification with the two primary political parties has declined, and the number of people ready to earmark even nominal monies in Internal Revenue Service tax returns for political campaigns has gone down significantly since the 1960s) is in an inverse relationship with the expansion of judicial rights—in the courts, at work, in the home, and at school. Rights guaranteed through the courts have become so expansive at the very time interest in traditional republican politics has declined.[46]

The waning of republican politics is associated with the decline of a civic sphere that defines the collective "will," because there is no longer any "collective" will. Indeed, issues of *agency* have replaced dedication to the *democratic* and republican process. Agency concerns the ability of the individual to act as an "initiatory" and "self-reliant" actor and to be an active participant in determining his or her life, including the determination of social, political, cultural, ethnic, religious, and economic ends.[47] In contrast with the past, no area of life today is beyond the potential reach of the web of the law.[48] And thus it is the dense web of legal rights and restraints which constitutes the mechanisms of "self-determination," not the civic sphere or the public square. The politics of rights is not about the politics of consent as such. In contrast to the shared space of the public square, the stress is on private space, on suburban space: "where space is relatively abundant, and public and economic limits are weak, as in the United States, community spatial patterns are very much the product, for the most part unintended, of millions of such private spatial choices."[49] And we can extend that notion of space from the geographic to the social, where through such agency we maintain our own social spaces linking up with others in ways that are independent of any bounded national republic. Such agency is linked to the increasingly "discrete" nature of the individual, as described by Simmel, and tied to different "circles" increasingly tangential from one another. This is all the more so in a world where the one compulsory membership of the nation-state, which enforced a republican arena and a common identity, is frayed.

The political scientist Galya Benarieh Ruffer called the myriad set of organizationally based rights and protections a "virtual citizenship," which in important respects is independent of national citizenship. Thus the rights and protections that accrue to an employee, for example, in an American corporation, public or private, are not a function of formal citizenship status (though, of course, the presumption is that the employee is at least a legal resident).[50] In this regard as well, the expansion of rights has diluted national citizenship, at least in its traditional republican sense.

The growing web of judicial rights in all kinds of organizational contexts makes agency possible. Agency of all kinds of individual actors generates adjudication of rights, interests, and the like. Thus agency reinforces the process of judicial rights which made it possible in the first place. The democratic process is not removed so much as it is contained. This is illustrated in the following example: the former president of Stanford University, Gerhard Casper, noted how Supreme Court justice Felix Frankfurter talked of the four freedoms of the university: the freedom to determine who may teach, what may be taught, how it shall be taught, and who shall be admitted for study. There was a time, Casper suggested, when the president of the university would consult with the different constituencies of the university—faculty, trustees, and students—on such issues. However, all those "four freedoms" are now subject to government regulations that constrain such a process of consent. Furthermore, there are a variety of committees, caucuses, and boards internal to the university, as well as administrative bodies and courts external to it, which can make the university legally accountable for its actions.[51] Thus what may have once been understood as a democratic process has become more multifaceted. The democratic process is contained within certain bounds, while agency for individuals has expanded.

These social and political changes away from traditional republicanism do not suggest that contemporary life is "displaced"; rather, the arrangement, the pattern of life, its spatial organization and tempo are being layered in different ways. "Place," in the larger sense of locating people spatially and temporally, has become interior to people in everyday life, privatized in the public sphere (which in turn shapes the form and content of that public life). But in a "spiritual rhythm expressed in the movement of life" people *place* themselves, orient themselves to the world, through parks, museums, diasporic ties, and sites of pilgrimage. Life has become, goes the truism, much more mobile. Indeed, movement is the norm in a country like the United States, as one observer wrote, making the incidental spaces we inhabit anchor us less and less. Until the mid-nineteenth century distance was defined by walking distance or the horse's lope. Neighboring medieval villages often had different weights and measures. Now we use terms such as "space-time compression" to indicate a world that acts in rationalized concert. Yet in this infinite expanse of time and space, but in a more individual and multifaceted way, it is in nature and the land that we find a "place of irreducible localness whose claims upon us cannot be annulled no matter how hard we try."[52] We still seek our *axis-mundi.*

Finding America's Place in the Global Landscape

Rarely has a country stood for more things to more people than America. And rarely has a country been made to stand for such wildly divergent things as America, for its images in the global imagination have rocked back and forth from paradise to foreboding wilderness, honest and open to hypocritical, redeemer nation to satanic incarnation. It is as if America sits on a pendulum, rocking back and forth between light and darkness.[53]

Over the past century we have similarly been witness to wildly divergent images of the United States on the global stage: as the savior of freedom, democracy, and the West in World Wars I and II and as a cynical and hypocritical imperialist in Latin America and Vietnam; as representing all that was good and decent and free about humankind and as a morally depraved and satanic empire assaulting the poor and the just. We have been witnesses to revolutions that invoke the Declaration of Independence as their inspiration and revolutions that were driven by anti-American passions.

There is little doubt that America's extraordinary presence in the world's mind has acted as a catalyst for immigration. America beckons, like the Statue of Liberty's torch. The linkages are woven through television, foreign aid programs, the letters and phone calls of family members who have already migrated, and television shows beamed around the world.[54] In some cases immigration may be driven by the promises of political freedom for many refugees; for others it is an economic draw of the *golden medinah,* the golden country, as Jewish immigrants from Eastern Europe at the turn of the twentieth century called it. But even economically driven migration has a cultural root: the *golden medinah* was a bundle of economic promise and political freedom, the ability to begin again—or so it was perceived. It was a country said to be moving forward, animated by the idea of progress and not fettered by the past. There is, and has been, a curious interplay of America having a mission in the world and its magnetic attraction to outsiders. Indeed, there is a certain dialectic here: in seeking to advance a universal mission, the county contributes itself as the "universal nation" and thus as a "nation of immigrants."

What is now raised as a question *outside* the United States, however, is to what extent the world, so preoccupied by America and so taken by American consumer culture, is itself being "Americanized." American culture is, as it were, immigrating to the rest of the world. The symbols of economic and cultural globalization—Coca-Cola, McDonald's, Nike, Hollywood, media, and television programming—tend to be identified with the United States. One could even argue that the United States is the epicenter of globalization, even, or especially, in the cultural sense of individual values, consumerism, and the increasing rationalization of life. Globalization, like America itself, generates a certain ambivalence: the consumer choices, the economic growth, and even the cultural possibilities are welcomed. On the other hand, the apparent effects of homogenization and of the flattening of local cultures which appears in the wake of globalization, as well as the economic dislocations, foster a definite trepidation. No wonder, in this context, McDonald's is attacked (quite literally) in France for, among other reasons, threatening local culinary culture.

A curious interplay is at work here in another sense as well, that of America as home to a polyglot of cultures, a global nation, while the world itself becomes home to all things American. Wide swaths of the world as a whole are becoming, piecemeal, privatized public spaces, where the intertwined moral and material, and metaphorical and literal, meanings of intellectual property apply. The dynamic movement, the restlessness, the kaleidoscope of American culture are at home abroad.

The Labyrinth of the Soul

CODA

ALL OF US, at some moment, have had a vision of our existence as something unique, untransferable and very precious," observed the writer Octavio Paz, adding, "Self-discovery is above all the realization that we are alone: it is the opening of an impalpable, transparent wall—that of our consciousness—between the world and ourselves."[1] The consciousness is, to put it another way, unique and bounded. We cannot explain in any material sense the "I"—why I am I, why I am, why this moment. We will perhaps get to the point where we will be able explain human actions and to attribute such actions to specific biological and environmental factors. But we cannot even begin to explain, in any positivistic sense, the I, being.

We can perceive the world and frame it, but only in material terms; only what is perceptible is real to us in any concrete sense. But this creates a deeply paradoxical situation. The "I" is unique, fixed in time and singular, irreproducible, and, in any rational sense, unexplainable. The material world is all that is real, yet the self, the "soul," the I, remains immaterial. Thus we "know" God, in all his forms, yet he is humanly unknowable.

The dilemma is not, historically at least, "Is there a God?"; the dilemma is that we cannot *know* God. In Hebrew God is referred to as "HaShem," the Name, unknowable, awesome in his absence.

History is a story, at heart, about the unfolding of this paradox, about this question of meaning, of the betwixt and between, the paradoxical character of humanity. It is an issue that has generally been referred to as the "material" and the "ideational," and much debate has revolved around the relationship between the material world and ideas, albeit phrased and argued in different ways.

The story of Genesis (meaning birth, creation, and formation) is about the creation of form, shape, and, by extension, knowledge. We can only know through categories, forms, words, and delineations. And yet, take that knowledge, sharpen it, multiply the categories, systemize it, make it more precise and detailed, rationalize it, and then the mystery is lost. Humankind can "know" only through form, including knowing God; yet God is the Name—formless, the All in All. Rationalization, the process at the heart of modern life, is the *perfection* of form—methodical, structured, calculated; not coincidently, it places God in the shadowlands. And so this dialectic, between form and formlessness, plays itself out. Genesis captures a profound dilemma: we need form to know, including knowing God; but knowledge, formed and classified, cannot capture God in his formlessness, his absence.[2]

Place is necessary in order to be oriented to the world. To state that we are Americans, Brazilians, or Botswanans is not only a geographic statement but also a moral statement about communal affiliations, ties, political beliefs, associations, and claims. And as a moral statement it is a form or category of knowledge, of knowing the world and one's place in it, literally and metaphorically.

The interplay of life as biological and physical fact and life as "being" is captured religiously and mythologically in Gen. 2:7: "Lord God formed man of the dust of the ground, and breathed into his nostrils the breath of life; and man became a living soul." The first human's name in the Genesis story, Adam, means "man" and "mankind": it is also the root in Hebrew for land or soil (*adama*). Physical and metaphysical, matter and antimatter, interpenetrate, fuse, and transport each other. The ground on which humans move and locate themselves locates them in a metaphysical universe. And so God in Gen. 2:8 gives "place" to Adam: "And the Lord God planted a garden eastward of Eden; and there he put the man whom he had formed." This place was also home, for when God banished Adam and Eve (for eating from the tree of knowledge!), they were sent into exile—hence the theme of "wandering" and being "lost." Ever since, the faithful have sought the Promised Land, the Garden, lush in its promise, the new Eden, the "fresh green breast of the new world." Exile, wandering, Great Migration, New Israel, America.

The Feeling of Being Here

The neurobiologist Antonio Damasio doubts that explaining *the* "consciousness problem" biologically—why we *are*—will ever be in our grasp. But he seeks to explain biologically what is self-evident to any human being, namely, that the "self" is, indeed, *self*-evident; it is experienced; it is. "What could be more difficult [than] to know how we know?" he asks, further adding, "What could be more dizzying than to realize that it is our having consciousness which makes possible and even inevitable our questions about consciousness?" Damasio thus seeks to explain not only how humans engender mental patterns, the "temporally and spatially integrated mental images of something-to-be-known," be it melodies or places, but also how a sense of self is engendered in the act of knowing. At base, he states that the neurobiology of consciousness is faced with two nested problems: first, how a "movie-in-the-brain" is generated, and, second, how the appearance of observing (and owning) the movie, while "watching the movie," is generated.[3]

In addressing these issues, Damasio makes a distinction between *core* consciousness and *extended* consciousness. Core consciousness concerns the processing of images in the given present, here and now. It does not picture the past, except for the instantaneous past, and cannot envisage the future. It is the "movie" without the appearance of watching the movie. There is no other place, and there is no past or future. Extended consciousness provides for a much more elaborate self in which the "past and the anticipated future are sensed along with the here and now in a sweeping vista as far-ranging as . . . an epic novel." Although extended consciousness may be present in nonhumans, it is at its most developed in humans. Nor is it an independent form of consciousness but is predicated on a core consciousness. Organisms without an extended consciousness "carry out" various operations and actions but are not aware of these operations and actions as they do not know of their existence as individuals—they are not aware that they are observing the "movie" or aware of being located in a certain point in time and space.[4]

It is the "extended consciousness"—the I is aware that it is here observing itself registering sensory information in a given present—that opens us up to the world by making us aware of time and space and by locating ourselves historically and spatially. This is why place is so central to humans. Place is the intersection where the "I" is; insofar as the "I" is *self*-conscious, it is conscious of the self as finite—in a stream of time and space which is infinite. The search for God is the search for place. The search for place is the search for God (or, more broadly, some metaphysical ordering of the infinite, potentially seamless, "chaotic" streaming of space and time). "Place," in other words, links and reconciles the present with a larger spatial and temporal schema.[5]

I am aware, in observing myself, that I am *here,* which means I am *not there*—spatially and temporally. Place is *here;* when I am *there,* I am "out of place." As humans we are constantly *locating* ourselves, always mediating between the finite moment we are in and the almost infinite vistas of memories and futures, utopias and dystopias we can imagine. If we just processed information in the ongoing present, "here" and "there," "now" and "then" would not exist. Thus "place" is a constant of the human condition. The search for place (or finding it) in the "cosmos" is the search for locating the self in the infinite universe of space and time.[6] The question of place, beyond territorial imperatives, is what distinguishes humans from other creatures. The temporal and spatial dimensions of place also indicate why the metaphysical dimension, large and small, from God to icons, from theology to the spatial tempo of funeral rites, is part of the definition of life.

A Grain of Sand

Whether the "self as soul" is or is not immortal is an issue we can put aside here. What is clear is that the "extended consciousness" makes possible the exploration across time and space of different possibilities. It is the individual "mind" that moves across space and time, but the "maps" are largely externally derived, just like language. Death marks the end of physical existence—the finiteness of the individual's ability to range in time and space. Yet, because of this ability to "explore" (through the mind) through time and space, one can soar to immortal heights. This, similarly, marks humanity; it also marks us socially and politically in that while attempting to transcend this reality we seek to "locate" ourselves spatially and temporally, as nations and as religions.

We can entertain the idea of God, or the metaphysical equivalent, because we are capable in our individual and collective imaginations of ranging across time and space. Place as the intersection of time and space reconciles the "here and now" with the "forever and everywhere." To "know" God is to be located in history and spatially (much like the Puritans, say, viewed themselves and their New Jerusalem as located at a certain point in history and space). God, however, is said to be omnipresent and omniscient; it is if space-time is indeed, in this context, a constant: God "is located and locates" infinitely, as if the "here and now" is "everywhere and forever" and the "everywhere and forever" is always "here and how"; as if the finite and the infinite are reconciled: "To see a World in a Grain of Sand, And a Heaven in a Wild Flower, Hold Infinity in the palm of your hand, And Eternity in an hour."[7] The belief in transcendent truths reconciles the human dilemma, the uncertainty of the human condition—it is an attempt to reconcile the material and ideal, the bounded physicality of human existence with the metaphysical "range" (across space and time) of human consciousness.

Contained in this the dilemma is the question, Why are *this moment* and *this space* that "I" (or you, or we, or they) occupy privileged? Why is the *present* present? In an extension of that question, Why should a given individual or society exist or be born into *this* moment and space and not *that* moment in the past or future?

The "I" knows itself as bounded and finite. I am aware that I sense myself as a discrete entity, alive, conscious, relating to others. That awareness is spatial and temporal. So the boundedness, the finiteness, of the self is relative to the spatial and the temporal dimensions. So the "I" understands itself as a function of locality—why here, why now? Because we can range over time and space, we *feel* the present moment as privileged (unlike, say, the cow who just is: the present is not "privileged"—the present simply is what is—there is no other moment such that the present can be constituted as privileged). But we feel the present moment as privileged only in the sense that we are "aware" of our finiteness (spatially and temporally), that we are located historically and geographically. Only the present is palpable in any positive or concrete sense, yet we act based on past events (the way we rely on language, for example, that preexists the moment) and act in terms of anticipating future developments. We are creatures that think in terms of cognitively mapping, plotting on the basis of where we came from and where we think we going, temporally as well as spatially.

We have to locate ourselves historically and spatially to make the present comprehensible, but the present is comprehensible only if it is extended "transhistorically," by making the moment eternal. When the moment is transcended, it is then that we become suprahistorical, as it were, that that moment becomes eternal, that the "now and here" become "forever and everywhere" and the "forever and everywhere" become "now and here"—in other words, the All in All.[8] Although the "suprahistorical" may be expressed most readily in religious terms, this is not exclusively the case: patriotism, nationalism, romantic love, and the arts all contain expressions of such felt epiphanies.

Place locates individuals and societies of people in a point in time and space and in doing so resolves—potentially—the (infinite) stream of space and time, by *orienting* human beings in a given society, culture, or civilization, *placing* them in the cosmos. This is the human understanding that resolves, or seeks to resolve, the physicality and "extended consciousness" of existence—to look backward and forward in time, inside and out in space, and to find in space-time a point, a place, in the cosmos. In doing so *every* point in space-time—past, present, future, here, and there—becomes significant or meaningful, part of an existential journey. (It is, indeed, like a journey, a walk, from point A to point B. One makes "progress" to point B, but that does not make point A any less "real," or privileged; so time or history is conceived insofar as time takes on a sacred and linear significance.) The sacred makes the finite, in time and space, infinite; and it makes explicable why time flows and the present is present. The forms of organizing space and time will vary enormously, of course, and mapping points in each case will be plotted differently whether we are talking of Western civilization, Confucian China, American Indians, or African societies, to cite a handful.

Consequently the relationship with nature, the land and the soil in their prosaic sense, represents the material, the here and now; but in its sacralization it reconciles the here and now with the everywhere and forever. The soil is not simply dirt, present; it conjures up the sacred, eternal. "America," rather than just dust, was a chosen place that organized time and space, a moment on the wings of God. "The earth is not a mere fragment of dead history," Thoreau wrote, "stratum upon stratum like the leaves of a book, to be studied by geologists and antiquaries chiefly, but living poetry like the leaves of a tree, which precedes flowers and fruit, not fossil earth, but a living earth."[9] Monuments play a similar role, making the here and now everywhere and forever, making the moment eternal. The "moment" of the monument is always privileged, always here and now.

The Poetry of Place

Art, in the broadest sense, is an amalgam of the objective and subjective, matter and antimatter. Out of clay and marble, mundane and inert, we try to capture transcendent emotions—love, patriotism, jealously, beauty, and passion. Like a Rodin sculpture—*La Danaïde* or *The Eternal Kiss,* for example—the wellspring of human emotions, of love and desire, emanates from the stone, the very grain, of the earth. It is the play between physicality and life: In the words of Rodin, "The human body is above all the mirror of the soul, and [there lies] its greatest beauty."[10]

What is it about aesthetic form which transcends time and space? What is in the clay that can represent nature, nation, desire, God? Form, after all, implies discreteness—inside and outside; here, not there; now, not then—which by definition is not transcendent. As humans, we cannot comprehend formlessness intellectually; yet people seek the All in All, which transcends the momentary, the finite, the fleeting, the transient that, by itself, does not locate us in the larger schema of things. The illumination of the sculptor, composer, poet, writer, or painter is to take form and show the infinite, to use form as a window to the All in All. Think of how cubism, for example, can combine movement, which takes place over time, and space, which is in the present tense; as the Russian artist Alexander Shevchenko wrote: "Cubism allows us to see and depict objects in greater relief, it allows us to facet them. On the one hand, this simplifies their form and, on the other, simultaneously, exposes it more fully . . . Cubism allows us to transfer planes so that we can see the object from several sides simultaneously without going around it . . . [Hence] Cubism also lets us see movement."[11]

Yet art is also a material artifact whose form can be "read" only in certain times and places. Thus we can "intellectualize" and analyze art, locate it like the Hudson River School as belonging to a certain period, suggest that it is the product of a certain *Zeitgeist,* and so on. When we analyze an event or object, we fix it historically and geographically—thus making it "mundane." By doing so the object is stripped of its awe—the notion that it transcends the moment. This is true in terms of political beliefs and myths, as in the case of Catholic icons on the eve of the Protestant Reformation—they suddenly became "worldly," mundane, objects of the here and now, simply *there.* Mystics who take vows of silence recognize the potential worldliness of any "form," including language. But, by definition, such an approach necessitates a withdrawal from the world.

America—indeed, any place—is a picture. It is a picture or a series of pictures depicting not only a corporeal reality but also a constant interplay of subject and object, of being located in a greater stream of space and time. In the undulations of preserved landscape, in surveying, urban planning, architecture, and parks, places reflect an aesthetic that melds object and subject. And so Thomas Jefferson was a landscape painter; the Hudson River School artists were statesmen.

At Home in Exile

The labyrinth, with its interconnecting passageways and blind alleys arranged in mazelike complexity, sets up an arduous journey with the sometimes illusory promise of getting "home," to the center. During the Middles Ages labyrinths were marked into the floors of cathedrals. Perhaps these labyrinths alluded to the seeking of an ever elusive God, who would reconcile finally the great puzzle of life.[12] The labyrinth contains the intricate relationship of time and space: if life is a journey (spatially and temporally), it is a journey with a place in mind, a place difficult to reach and ever receding.

This is the paradox and beauty of life, of humanity—ever seeking the place, the moment when we finally arrive, the "fresh green breast," the warmth, the love, fertility, Eden, nature, the sweet garden of paradise, the moment eternal.

NOTES

INTRODUCTION: TERRA FIRMA

1. The notion of "home" and place is discussed in Steven Grosby, "Territoriality: The Transcendental, Primordial Feature of Modern Societies," *Nations and Nationalism* 1, no. 2 (1995): 150.

2. James Elroy Flecker, "Brumana, " in *The Collected Poems of James Elroy Flecker,* ed. J. C. Squire (New York: Knopf, 1921), 181.

3. The geographer Yi-Fu Tuan notes, in *Cosmos and Hearth: A Cosmopolite's Viewpoint* (Minneapolis: University of Minnesota Press, 1996), the contesting forces—and attractions—of hearth *and* cosmopolitanism; hearth is nurturing but may be confining, and the cosmopolitan life, though liberating, can be bewildering. These themes will be touched upon, if in a different sense, in Chapter 5. For a general discussion on space, see also Yi-Fu Tuan, *Space and Place: The Perspective of Experience* (Minneapolis: University of Minnesota Press, 1977).

4. Stephen Cornell, "The Transformations of the Tribe: Organization and Self-Concept in Native American Ethnicities," *Ethnic and Racial Studies* 11, no. 1 (1988): 29–31.

5. Ernest Wallace and Hoebl E. Adamson, *The Comanches: Lords of the South Plains* (Norman: University of Oklahoma Press, 1952), 22, quoted in Cornell, "Transformations of the Tribe," 30.

6. Cornell, "Transformations of the Tribe," 32; Cornell also cites Imre Sutton, *Indian Land Tenure* (New York: Clearwater, 1975), 192–93.

7. Tibor Szamuely, *The Russian Tradition* (New York: McGraw-Hill, 1974), 12, 23–24, 37. See also Adam B. Ulam, *Expansion and Coexistence* (New York: Praeger, 1974), 4.

8. Richard Pipes, "Russia's Mission, America's Destiny," in his *U.S.-Soviet Relations in the Era of Detente* (Boulder, Colo.: Westview Press, 1981), 5.

9. Mark Mancall, "The Ch'ing Tribute System: An Interpretive Essay," in *The Chinese World Order,* ed. John K. Fairbank (Cambridge: Harvard University Press, 1968), 63–64.

10. John K. Fairbank, "A Preliminary Framework," in his *Chinese World Order,* 9.

11. Ibid., 1–7.

12. Joseph R. Levenson, *Confucian China and Its Modern Fate: A Trilogy* (Berkeley: University of California Press, 1968), 3:104–5.

13. Daniel Boorstin, *The Discoverers* (New York: Vintage, 1985), 104–5.

14. Steven Grosby, "Sociological Implications of the Distinction between 'Locality' and Extended 'Territory' with Particular Reference to the Old Testament," *Social Compass* 40, no. 2 (1993): 185.

15. Migration, Rogers Brubaker has written, "did not engage so directly the vital interests of ancient or medieval personal polities since rule in these settings was exercised over particular sets of persons, not over territories: mere presence did not entail political, administrative, or legal inclusion." Furthermore, "since jurisdiction depended on the personal status of the agent rather than the spatial coordinates of the action, migration was less consequential." Rogers Brubaker, *Citizenship and Nationhood in France and Germany* (Cambridge: Harvard University Press, 1992), 25.

16. See Michael Walzer, *The Revolution of the Saints: A Study in the Origins of Radical Politics* (Cambridge: Harvard University Press, 1965).

17. *Treaty between the United States of America and the Navajo Tribe of Indians with a Record of the Discussions That Led to Its Signing,* with an introduction by Martin A. Link (Las Vegas: KC Publications, 1968). The Navajos had been forcibly removed in 1863 by troops commanded by Kit Carson, the frontiersman, Indian agent, trapper, and soldier who symbolized for many the westward expansion. The negotiations between the Navajos and Sherman in the spring of 1868 took some three days, after which a treaty was signed allowing the Navajos to return to their native lands. Emphasis in quotation added.

18. The use of territoriality as a central element in demarcating a political and social space and republican government is associated with the acceptance by such entities of other *like* states, whether they are referred to as civic polities, liberal democracies, or nation-states. In a people's claim to the right of self-government (and self-determination), there is an implicit or even explicit premise that that right is a universal principle. The nation is portrayed as not only a legal personality but, so to speak, a social being as well, "a generalized self." Just as international law recognizes states as legal persons with rights and duties, so nation-states recognize that there is a plurality of actors or states, defined and guided by the same norms, on the international stage. The nation-state embodies, at least imputedly, "a general will": as such it recognizes other "wills." In claiming national *self*-determination, "other-determination" is recog-

nized as well—within limits. Are these "others" of a republican nature? Do they recognize their own boundedness and territorial limits? In this context, the findings that liberal democracies tend not to go to war with each other is telling. In contrast, in the postwar Stalinist Soviet Union, "states" as such were secondary. The image of the actors in the world arena was of the two "world systems," socialist and capitalist. Intrabloc homogeneity and coordination were postulated. Territorial boundaries were tactical in that they, ideologically speaking, had no permanent significance. Maoist China was similar in that state boundaries were a temporary phenomenon; indeed, the language of diplomacy was quite consciously of "people-to-people" relations rather than state-to-state relations. Mao's internationalist stance rejected ideas of corporate nationhood and bounded territoriality. As in the case of imperial Russia and China, no "natural borders" were recognized. Recognition of territoriality, in this regard, has been associated with internal reform: again, the idea of self-government has international as well as domestic implications. See C. A. W. Manning, *The Nature of International Society* (New York: Wiley, 1962), 8–16, and William Zimmerman, *Soviet Perspectives on International Relations* (Princeton: Princeton University Press, 1969), chaps. 1–3.

19. Joseph R. Strayer, *On the Medieval Origins of the Modern State* (Princeton: Princeton University Press, 1970).

20. See Linda Kay Davidson and Maryjane Dunn-Wood, *Pilgrimage in the Middle Ages* (New York: Garland Publishing, 1993).

21. Grosby, "Sociological Implications," 187, 189; emphasis in original. Grosby discusses how, with "territorialization," the concept of the sacred does not require the believer to be at specific sites (185).

22. Cornell, "Transformations of Tribe," 39–40. Cornell cites Raymond DeMallie, "Pine Ridge Economy: Cultural and Historical Perspectives," in *American Indian Economic Development*, ed. Sam Stanley (The Hague: Mouton, 1978), and Albert J. Wahrhaftig and Jane Lukens-Wahrhaftig, "The Thrice Powerless: Cherokee Indians in Oklahoma," in *The Anthropology of Power*, ed. Raymond D. Fogelson and Richard N. Adams (New York: Academic Press, 1977). Interestingly, it has been argued that immigration can act as a mechanism to break kinship connections and produce national or supratribal identities. Italian immigrants to the United States at the turn of the twentieth century would leave Italy as, for example, Sicilians but evolve into considering themselves "Italians," as in Italian American, once in America. The self-identification of a Navajo, say, as part of a larger category of Native Americans or American Indians is associated with migration into urban areas.

23. See the discussion of this debate in Anthony D. Smith, *The Nation in History: Historiographical Debates about Ethnicity and Nationalism* (Hanover, N.H.: University Press of New England, 2000), chap. 2.

24. Susan Reynolds, *Kingdoms and Communities in Western Europe, 900–1300*, 2d ed. (Oxford: Clarendon Press, 1997).

25. Ibid., lxiv. See also Strayer, *Medieval Origins of the Modern State.*

26. Robert David Sack, *Human Territoriality: Its Theory and History* (Cambridge: Cambridge University Press, 1986), 129–30.

27. Robert D. Putnam, *Bowling Alone: The Collapse and Revival of American Community* (New York: Simon and Schuster, 2000).

28. See David Jacobson, *Rights across Borders: Immigration and the Decline of Citizenship* (Baltimore: Johns Hopkins University Press, 1986).

29. This not to say, of course, that other factors do not play a role as well.

30. See discussion in Stanford M. Lyman, "Interstate Relations and the Sociological Imagination Revisited: From Social Distance to Territoriality," *Sociological Inquiry* 65, no. 2 (1995): 125–42. The part that follows draws on David Jacobson, "New Frontiers: Territory, Social Spaces, and the State," *Sociological Forum* 12, no. 1 (1997): 121–33.

31. Erving Goffman, *Relations in Public: Microstudies of the Public Order* (New York: Basic Books, 1971). Goffman explains the notion of "marker" on p. 41.

32. Lyman, "Interstate Relations and the Sociological Imagination," 126.

33. Edward Alsworth Ross, *Social Control* (New York: Macmillan, 1908), 1; cited in Goffman, *Relations in Public,* 6.

34. Georg Simmel, *The Sociology of Georg Simmel,* trans. Kurt H. Wolff (Glencoe, Ill.: Free Press, 1950), 322.

35. Ibid., 315.

36. See Ezra Vogel, *Canton under Communism: Programs and Politics in a Provincial Capital, 1949–1968* (Cambridge: Harvard University Press, 1969).

37. In Emile Durkheim's terms, the individual "depends upon [others] to the very extent he is distinguished from them," and these sets of relationships are inherently moral. See Durkheim, *The Division of Labor in Society,* trans. W. D. Halls (1893; New York: Free Press, 1984), 172.

38. Lyman, "Interstate Relations and the Sociological Imagination," 132.

39. Michael Walzer, *Spheres of Justice: A Defense of Pluralism and Equality* (New York: Basic Books), 62–63.

40. See discussion in Christian Joppke, "Multiculturalism and Immigration: A Comparison of the United States, Germany, and Britain," in *The Immigration Reader: America in Comparative Perspective,* ed. David Jacobson (Oxford: Blackwell Publishers, 1998), 285–319. In the early and mid-1990s, as issues of multiculturalism became increasingly debated, a curious ambivalence apparently characterized the public. From surveys of the time it appears that there was simultaneously unease over increasing ethnic and racial divisions and yet a positive view of a multiethnic society. Polls indicated an overwhelming sense among Americans that racial and ethnic tensions had increased, and yet, "on balance," Americans believed that increasing numbers of people from many different ethnic groups, races, and nationalities make the United States "a better place to live." Similarly, only 28 percent of survey respondents said that they are "Americans

first and a member of an ethnic group second." They are more likely to call themselves a "hyphenated" American if those categories are the only alternatives suggested. See respectively Washington Post poll for survey ended October 6, 1995, of 684 respondents (with an oversample of 279 blacks but with results weighted to represent national adult population); Princeton Survey Research Associates for survey ended January 15, 1996, with sample of 1,206 national registered voters; and Gallup Organization, World Values Study Group 1990–1993 (survey ended June 30, 1990), with national adult sample of 1,839. See *Public Opinion Online,* available through Westlaw, 2000.

41. See Jacobson, *Rights across Borders,* chap. 7.

42. Goffman, *Relations in Public,* chap. 2.

43. See essays in C. Neal Tate and Torbjörn Vallinder, *The Global Expansion of Judicial Power* (New York: New York University Press, 1995), and Martin Shapiro and Alec Stone, "The New Constitutional Politics of Europe," *Comparative Political Studies* 26, no. 4 (1994): 397.

44. Another element here is what Durkheim termed the growing importance (with the growing division of labor) of restitutive sanctions (civil law) and the concomitant diminution of repressive sanctions (criminal law) as the "collective conscience" itself contracts. Restitutive sanctions are about personal space—of character, of property. See Durkheim, *Division of Labor in Society.*

45. Goffman, *Relations in Public,* 6.

46. Mircea Eliade, *The Sacred and the Profane: The Nature of Religion* (New York: Harcourt, Brace and World, 1957), 57.

47. Goffman, *Relations in Public,* 60–61.

48. Ibid., 122.

49. See Hannah Arendt, *The Human Condition* (Chicago: University of Chicago Press, 1957).

50. One prominent critique of the traditional history emphasizing the role of Puritans in American culture is Jack P. Greene, *Pursuits of Happiness: The Social Development of Early Modern British Colonies and the Formation of American Culture* (Chapel Hill: University of North Carolina Press, 1988); quoted text on p. 38. However, his book does not contradict what is said here and in Chapter 1 and actually highlights certain salient points: the Puritans' special sense of mission (which was largely absent in the other colonies); the commercial orientations of other colonies; and the turning away from old England in New England (which Greene notes as contrary to the older historiography on the Puritans, which that claimed New England reproduced old England). See in Greene, especially pp. 28, 35–53, 97–100, 205, 208. The distinctions between New England and the other colonies noted here are generally accepted by revisionist historians such as Greene (as well as others) who in the last two decades or so have criticized, fairly, the attempt to "reduce" America's colonial origins to Puritan New England. Theirs is an effort to counter a theme in American historical writing, one that gives cen-

ter stage to the Puritans and Pilgrims, which goes back to the nineteenth century. Conversely, the revisionist historians have stressed the important role of other colonies in the early seventeenth century, such as Virginia, Pennsylvania, and the Carolinas. That revisionist argument will not be contested or even contemplated here, nor is anything like a comprehensive history of colonial America in this period attempted in Chapter 1.

ONE AN AMERICAN EDEN

1. Avihu Zakai, *Exile and Kingdom: History and Apocalypse in the Puritan Migration to America* (Cambridge: Cambridge University Press, 1992), 7.

2. The quotations are, respectively, from Shakespeare, *Richard II,* Act 2, Scene 1, and from William Blake, "Milton," in *William Blake: The Complete Poems,* ed. Alicia Ostriker (New York: Penguin Books, 1977), 514.

3. See the detailed description in Zakai, *Exile and Kingdom,* and Perry Miller and Thomas H. Johnson, "General Introduction," in *The Puritans,* ed. Perry Miller and Thomas H. Johnson (New York: American Book Co., 1938), 6.

4. Zakai, *Exile and Kingdom,* 64. Quotes are from a letter from John Winthrop to a friend in 1629 in the *Winthrop Papers,* vol. 2, *1623–1630* (Boston: Massachusetts Historical Society, 1931), 122.

5. John Winthrop to his wife, May 15, 1629, in the *Winthrop Papers,* 2:91–92.

6. Robert Ryece to John Winthrop, 1629, in the *Winthrop Papers,* 2:129.

7. Thomas Paine, "Thoughts on the Present State of American Affairs," in *The American Foreign Policy,* ed. Ernest May (New York: George Braziller, 1963), 27.

8. See Ernest Lee Tuveson, *Redeemer Nation: The Idea of America's Millennial Role* (Chicago: University of Chicago Press, 1968), 24–25, 45–46, 138–39, 142.

9. Robert Cushman, "Reasons and Considerations Touching the Lawfulness of Removing Out of England into the Parts of America," in *The Story of the Pilgrim Fathers, 1606–1623 A.D.; as told by Themselves, their Friends and their Enemies,* ed. Edward Arber (Boston: Houghton, Mifflin & Co., 1897), 497.

10. Zakai, *Exile and Kingdom,* 8.

11. I use the term *soul* here in the sense of Durkheim's use of the term, as described by Karen Fields: "likeness" of a community, such as "the Americans," is not spontaneously given but "stems from an intellectual process of abstraction"; likeness is an "invisible yet shared substance—*soul*—of which emblematic markings are merely an index, the outward and visible sign of an inward and spiritual thing." That describes the *felt* qualities of the "soul," but to be of social and even political significance this "invisible substance" acquires substance through doing, that is, through ritual acts and symbols. As Durkheim suggested in his *Elementary Forms of Religious Life,* trans. Joseph Ward Swain (New York: Free Press, 1965), the individual breaks out of him- or herself through symbols that engage (or disengage) others in the same collective body. See

Karen E. Fields, "Durkheim and the Idea of the Soul," *Theory and Society* 25 (1996):193–203; quote from p. 195.

12. Thomas Tillam, "Uppon the first sight of New-England June 29, 1638," in *Seventeenth-Century American Poetry,* ed. Harrison T. Meserole (Garden City, N.Y.: Anchor Books, 1968).

13. Perry Miller and Thomas H. Johnson, "History," in Miller and Johnson, *The Puritans,* 85–90.

14. William Bradford, *History of Plymouth Plantation, 1606–1646,* ed. William T. Davis (New York: Charles Scribner's Sons, 1909), 97.

15. Cotton was a strict believer in the Puritan theocracy: he was responsible for expelling Roger Williams and Anne Hutchinson, both of whom had promoted a more tolerant, latitudinarian Protestantism and who were ready to accept other religions in their midst.

16. The quotations from this and the following three paragraphs are from John Cotton, *God's Promise to His Plantations* (London: Printed by William Jones for John Bellamy, and to be sold at the three Golden Lyons by the Royal Exchange, 1634; reprinted at Boston in New-England, by Samuel Green, and to be sold by John Usher. Anno. 1686), 1–20; emphasis and spelling in original.

17. See "Arguments for the Plantation of New England," dated 1629, probably by John Winthrop, in the *Winthrop Papers,* 2:106–45.

18. See Michael Walzer, *The Revolution of the Saints: A Study in the Origins of Radical Politics* (Cambridge: Harvard University Press, 1965).

19. The nuanced, legal character of the Covenant is clearly to be distinguished from liberal notions of the social contract; see Michael Sandel, *Democracy's Discontent: America in Search of a Public Philosophy* (Cambridge: Harvard University Press, 1996).

20. It is significant that government in New England, in the absence of kinship as the principle of governance, took the form of geographic representation. See Perry Miller, *Errand into the Wilderness* (Cambridge: Harvard University Press, Belknap Press, 1956), 37–38. See also the discussion in the following chapter on geography and representation.

21. This quotation from Winthrop and those in the following four paragraphs are from Winthrop, "A Modell of Christian Charity," in *Winthrop Papers,* 2:282–95; emphasis added.

22. Miller, *Errand into the Wilderness,* 89

23. Cotton commands, similarly, to go forth with a "public spirit," *God's Promise to His Plantations,* 18; emphasis added.

24. Miller, *Errand into the Wilderness,* 12.

25. Ibid., 6.

26. "Ryece to Winthrop," *Winthrop Papers,* 2:130–32.

27. Miller, *Errand into the Wilderness,* 7–8. Miller cites Increase Mather's *Brief History of the Warr With the Indians.*

28. See Steven Grosby, "Sociological Implications of the Distinction between 'Locality' and Extended 'Territory' with Particular Reference to the Old Testament," *Social Compass* 40, no. 2 (1993), and Mircea Eliade, *The Sacred and the Profane: The Nature of Religion* (New York: Harcourt, Brace and World, 1957).

29. Cotton, *God's Promise to his Plantations,* 17; emphasis in original.

30. Ibid., 6.

31. Grosby, "Sociological Implications"; quotation on p.189.

32. Ibid.

33. Kenneth A. Lockridge, *A New England Town, the First Hundred Years: Dedham, Massachusetts, 1636–1736* (New York: Norton, 1970), 5.

34. See Perry Miller and Thomas H. Johnson, "Theory of State and Society," in Miller and Johnson, *The Puritans,* 183. The notion of the "umpire" state is intriguing, especially with regard to present-day contemporary developments and the rise of the judiciary, which, as I suggest in Chapter 5, may well be in opposition to the classical republican state.

35. Miller, *Errand into the Wilderness,* 21–30; quotation, noted in Miller, from Robert Ashton, ed., *John Robinson's Works* (Boston, 1851), 3:42–43.

36. Zakai, *Exile and Kingdom,* 244; quotation, noted in Zakai, from "Certain Proposals Made by Lord Say, Lord Brooke, and Other Persons of Quality, as Conditions of Their Removing to New England, with Answers Thereto," in Thomas Hutchinson, *The History of the Colony and Province of Massachusetts Bay,* ed. Lawrence Mayo (Cambridge: Harvard University Press, 1936), 1:413.

37. See Harold Laski, *Foundations of Sovereignty* (1921; Freeport, N.Y.: Books for Libraries Press, 1968), and F. H. Hinsley, *Sovereignty* (London: Watts, 1966).

38. See Ernst Troeltsch, *Protestantism and Progress; A Historical Study of the Relation of Protestantism to the Modern World,* trans. W. Montgomery (Boston: Beacon Press, 1958).

39. See Walzer, *Revolution of the Saints*; J. G. A. Pocock, *The Machiavellian Moment* (Princeton: Princeton University Press, 1975); and David Jacobson, "State and Society in a World Unbound," in *Public Rights, Public Rules: Constituting Citizens in the World Polity and National Polity,* ed. Connie L. McNeely (New York: Garland, 1998), 41–58.

40. Rogers Smith, *Civic Ideals: Conflicting Visions of Citizenship in U.S. History* (New Haven: Yale University Press, 1999).

41. David Jacobson, "Protestantism, Authoritarianism, and Democracy: A Comparison of the Netherlands, the United States, and South Africa," *Religion* 17 (July 1987): 275–301.

42. Roderick Nash, *Wilderness and the American Mind,* rev. ed. (New Haven: Yale University Press, 1973), 35–37.

43. Leo Marx, *The Machine in the Garden: Technology and the Pastoral Ideal in America* (New York: Oxford University Press, 1964), 41–42. He quotes from Bradford's *History of Plymouth Plantation,* 94–96.

44. Wrote John Winthrop, "[God] carried his people into the wildernesse, and made them forgett the fleshpottes of Egipt which was some pinch to them indeed, but he disposed it to their good, in the ende Deut: 8.3:16," in his "Arguments for the Plantation of New England," in *Winthrop Papers,* 2:136.

45. See George Huntston Williams, *Wilderness and Paradise in Christian Thought* (New York: Harper, 1962).

46. Miller, *Errand into the Wilderness.*

47. Winthrop, "Arguments for the Plantation of New England," 139, 118.

48. Orest Ranum, "The Refuge of Intimacy," in *A History of Private Life,* ed. Philippe Aries and Georges Duby (Cambridge: Harvard University Press, Belknap Press, 1989), 3:216. See also Belden C. Lane, *Landscapes of the Sacred: Geography and Narrative in American Spirituality* (New York: Paulist Press, 1988), 104 [on village layout].

49. Francis Higginson, *New-Englands Plantation Or, A Short and True Description of the Commodities and Discommodities of that Countrey* (London, Printed by T. C. and R. C. for Michael Sparks, dwelling at the Signe of the Blew Bible in Green Arbor in the little Old Bailey, 1630). Ironically, Higginson died from fever a year after writing these words.

50. Bradford, *History of Plymouth Plantation,* 47.

51. See Perry Miller and Thomas H. Johnson, "This World and the Next," in Miller and Johnson, *The Puritans,* 287, and Max Weber, *The Protestant Ethic and the Spirit of Capitalism,* trans. Talcott Parsons (New York: Scribner's, 1976), 154.

52. See definitions of *space* and *time,* respectively, in the *Oxford English Dictionary.*

53. Daniel J. Boorstin, *The Americans: The National Experience* (New York: Vintage Books, 1965), v.

54. John Winthrop, "Objections Answered: First Draft," in *Winthrop Papers,* 2:136.

55. Quoted in Lane, *Landscapes of the Sacred,* 104–5.

56. Perry Miller and Thomas H. Johnson, "Science," in Miller and Johnson, *The Puritans,* 730–34.

57. Weber, *Protestant Ethic.*

58. Bradford, *History of Plymouth Plantation,* 46–47.

59. Sacvan Bercovitch, in his forward to Charles M. Segal and David C. Stineback, *Puritans, Indians, and Manifest Destiny* (New York: G. P. Putnam's Sons, 1977), 17.

60. See Segal and Stineback, *Puritans, Indians, and Manifest Destiny,* 33.

61. Cotton, *God's Promise to his Plantations,* 19; emphasis in original.

62. Segal and Stineback, *Puritans, Indians, and Manifest Destiny,* 28.

63. Winthrop, "Arguments for the Plantation of New England," 120.

64. Cotton, *God's Promise to his Plantations,* 4–5.

65. Winthrop, "Arguments for the Plantation of New England," 120.

66. Segal and Stineback, *Puritans, Indians, and Manifest Destiny,* 27–30, 219–25.

67. Miller and Johnson, "Theory of State and Society," 247.

TWO SURVEYING THE LANDSCAPE

1. Edward L. Ayers and Peter S. Onuf, introduction to Edward L. Ayers, Patricia Nelson Limerick, Stephen Nissenbaum, and Peters S. Onuf, eds., *All Over the Map: Rethinking American Regions* (Baltimore: Johns Hopkins University Press, 1996), 1–2.

2. Individual states, such as Massachusetts, had state churches until as late as 1833. See Evarts B. Greene, *Religion and the State: The Making and Testing of an American Tradition* (New York: New York University Press, 1941), 92–93.

3. James McPherson, *Abraham Lincoln and the Second American Revolution* (New York: Oxford University Press, 1991), viii.

4. Daniel H. Deudney, "The Philadelphian System: Sovereignty, Arms Control, and Balance of Power in the American States-Union circa 1787–1861," *International Organization* 49, no. 2 (1995): 191–228.

5. See, generally, Merrill D. Peterson, *The Jefferson Image in the American Mind* (New York: Oxford University Press, 1962).

6. Henry Nash Smith, *Virgin Land: The American West as Symbol and Myth* (Cambridge: Harvard University Press, 1978), 133–54.

7. Peterson, *Jefferson Image in the American Mind,* 194.

8. Thongchai Winichakul, "Maps and the Formation of the Geo-Body of Siam," in *Asian Forms of the Nation,* ed. Stein Tonnesson and Hans Antilov (Richmond, Surrey, England: Curzon, 1996), 68–69.

9. See, for example, Rogers Smith, *Civic Ideals: Conflicting Visions of Citizenship in U.S. History* (New Haven: Yale University Press, 1999).

10. See Madison on democracy in *The Federalist,* no. 10. The edition referred to here is that of Clinton Rossiter, ed., *The Federalist Papers* (New York: New American Library, 1961), 77–84.

11. Roderick Nash, *Wilderness and the American Mind,* rev. ed. (New Haven: Yale University Press, 1973).

12. Charles A. Miller, *Jefferson and Nature: An Interpretation* (Baltimore: Johns Hopkins University Press, 1993); emphasis added.

13. Henry Van Dyke, "America For Me," in *The Best Loved Poems of the American People,* ed. Hazel Felleman (New York: Doubleday and Co., 1936), 424.

14. Rossiter, *Federalist Papers,* 37–41.

15. Thomas Jefferson, "First Inaugural Address, March 4, 1801," in *The Portable Jefferson,* ed. Merrill D. Peterson (New York: Viking, 1975), 290–93.

16. See David Noble, *Historians against History: The Frontier Thesis and the National Covenant in American Historical Writing since 1830* (Minneapolis: University of Minnesota Press, 1965), and Felix Gilbert, *To the Farewell Address: Ideas of Early American Foreign Policy* (Princeton: Princeton University Press, 1961).

17. George Thomas, "U.S. Discourse and Strategies in the New World Order," in *Old Nations, New World: Conceptions of World Order*, ed. David Jacobson (Boulder, Colo.: Westview Press, 1994), 152.

18. See Ernest Lee Tuveson, *Redeemer Nation: The Idea of America's Millennial Role* (Chicago: University of Chicago Press, 1968), 124.

19. Gilbert, *To the Farewell Address*, 4.

20. Noble, *Historians against History*, 3.

21. More generally, Protestant notions of equality, unmediated relationship with God, and congregationalism (with regard to voluntary and locally organized communities and churches) are present in the background of American republican self-description, if implicitly. The notion of America as an "asylum" carries over from the Puritans as well; as Paine writes in *Common Sense* (1776; New York: Bobbs-Merrill, 1953), 34: "O ye that love mankind! Ye that dare oppose, not only the tyranny, but the tyrant, stand forth! Every spot of the old world is overrun with oppression. Freedom hath been hunted round the globe. Asia and Africa, have long expelled her. Europe regards her like a stranger, and England hath given her warning to depart. O! receive the fugitive, and prepare in time *an asylum for mankind*" (emphasis added).

22. For Paine, the true word of God was in nature, not the Bible. See Thomas Paine, *The Age of Reason*, ed. Alburey Castell (New York: Liberal Arts Press, 1948).

23. Thomas Paine, *Rights of Man* (New York: Heritage Press, 1961), 38. The discussion on Paine and nature here has benefited from Jack Fruchtman Jr., *Thomas Paine and the Religion of Nature* (Baltimore: Johns Hopkins University Press, 1993).

24. Paine, *Rights of Man*, 51.

25. Paine, *Common Sense*, 10.

26. Paine, *Rights of Man*, 16.

27. Fruchtman, *Thomas Paine and the Religion of Nature*, 22, 41.

28. Paine, *Rights of Man*, 156.

29. Ibid., 261; see discussion in Fruchtman, *Thomas Paine and the Religion of Nature*, 51–52.

30. Paine, *Common Sense*, 11; emphasis in original.

31. Ibid., 5.

32. George Cary Eggleston, ed., *American War Ballads and Lyrics* (Great Neck, N.Y: Granger Book Co., 1889), 23–24; first published in the *Pennsylvania Magazine* in 1775. Emphasis in the original.

33. See Paine, *The Age of Reason*.

34. Paine, *Common Sense*, 5.

35. Hence a liberal attitude to immigration is explicit or implicit in such approaches, since peoplehood or nationhood arises through association and deliberation.

36. Thomas Paine, "An Answer To A Friend Regarding The Age Of Reason," also entitled "To Anonymous"; published in *The Prospect,* April 12, 1804.

37. Thomas Paine, "A letter to the Hon. Thomas Erskine, on the Prosecution of Thomas Williams for publishing the Age of Reason. By Thomas Paine, Author of Common Sense, Rights of Man, etc. With his discourse at the Society of the Theophilanthropists, Paris: Printed for the Author." This pamphlet was printed by Barrois's English press in Paris, September 1797; see also description in Leo Marx, *The Machine in the Garden: Technology and the Pastoral Ideal in America* (New York: Oxford University Press, 1964), 96.

38. On "constitutional invention" and Paine, see Fruchtman, *Thomas Paine and the Religion of Nature,* chap. 8.

39. Steven Grosby, "Territoriality: The Transcendental, Primordial Features of Modern Societies," *Nations and Nationalism* 1, no. 2 (1995):146.

40. Thomas Jefferson, *Notes on the State of Virginia,* ed. William Peden (Chapel Hill, N.C.: Institute of Early American History and Culture, 1955), 4.

41. Ibid., 10, 19, 25, 163. See also Jefferson on nationalism (64).

42. Quoted in Lewis Lord, "A 'Canine Appetite' for Books: How Jefferson Fathered the 'Nation's Library,'" *U.S. News and World Report,* May 1, 2000, 58.

43. J. Hector St. John Crèvecoeur, *Letters from an American Farmer* (1782; Gloucester, Mass.: Peter Smith, 1968), 82, 72–73, 80.

44. The essentially moral quality of capitalism, as portrayed at the time, is also evident in Benjamin Franklin's famous maxims. For Franklin the accumulation of capital was a moral issue, a happy outcome of "moral perfection." Among the virtues to be realized were order, resolution, frugality ("waste nothing"), and industry ("always be employed in something useful"). When a person possessed the virtues of industry and frugality, wealth was assured. The virtues were in themselves defined as wealth. What could be more descriptive of the capitalist (at least at the outset of capitalism, the point of greatest interest also to Max Weber) than Franklin's maxim that the process by which one accumulates wealth is more important than wealth itself! The process of accumulating wealth was sharply contrasted with simple greed or consumerism. The rational postponement of immediate gratification permeates Franklin's outlook: "Men have been drawn into . . . an overweening desire for sudden wealth . . . while the rational and almost certain methods of acquiring riches by industry and frugality are neglected or forgotten." See George L. Rogers, *Benjamin Franklin's The Art of Virtue: His Formula for Successful Living* (Minnesota: Acorn Publishing, 1986), 40–44, 159–71.

45. See Damian Thompson, *The End of Time* (New York: Vintage, 1996).

46. Grosby, "Territoriality," 146–47 (emphasis in the original); in a historical vein,

see Gordon S. Wood, *The Creation of the American Republic, 1776–1787* (Chapel Hill, N.C.: Institute of Early American History and Culture, 1969), 259–91.

47. Marx, *Machine in the Garden,* 73–74.

48. See ibid. on rural population numbers, 115.

49. See Peterson, *Jefferson Image in the American Mind,* 70.

50. Miller, *Jefferson and Nature,*1–10. F. Scott Fitzgerald's phrase appears in his *The Great Gatsby* (New York: Cambridge University Press, 1991), 140.

51. Thomas Jefferson, "Letter to James Madison, December 20, 1787," in *The Portable Jefferson,* 432.

52. Jefferson, *Notes on the State of Virginia,* 164–65 (emphasis added).

53. John Locke, *Second Treatise on Government,* ed. C. B. Macpherson (1690; Indianapolis: Hackett Publishing, 1980), 29. See also Leo Marx, *Machine in the Garden,* 120.

54. Smith, *Virgin Land,* 128

55. Thomas Jefferson, "Letter to James Madison, October 28, 1785," in *The Portable Jefferson,* 396–97.

56. Jefferson, *Notes on the State of Virginia,* 174–75; emphasis added.

57. See discussion in Miller, *Jefferson and Nature,* 205–11. The quotation from John Adams appears on 207 n; Miller cites Morton White, *The Philosophy of the American Revolution* (New York: Oxford University Press, 1978), 265–66. Jefferson's letter, to Horatio G. Spafford, March 17, 1814, is quoted in part in Miller, *Jefferson and Nature,* 210–11. As Miller observes (211), many urban dwellers were, in fact, supportive of the Jeffersonian republicans.

58. Crèvecoeur, *Letters from an American Farmer,* 30, 46, 48–50.

59. Mason is quoted in Michael Sandel, *Democracy's Discontent: America in Search of a Public Philosophy* (Cambridge: Harvard University Press, Belknap Press, 1996), 126.

60. Crèvecoeur, *Letters from an American Farmer,* 52–61.

61. Miller, *Jefferson and Nature,* 208–16.

62. Smith, *Virgin Land,* 126.

63. Jefferson, "Letter to James Madison" (1787).

64. See, in general, Miller, *Jefferson and Nature,* 158–61.

65. See Christian G. Fritz, "Constitution Making in Nineteenth-Century American West," in *Law for the Elephant, Law for the Beaver: Essays in the Legal History of the North American West,* ed. John McLaren, Hamar Foster, and Chet Orloff (Pasadena, Calif.: Ninth Judicial Circuit Historical Society, 1992), 292–320.

66. John Boli, "Human Rights or State Expansion? Cross-National Definitions of Constitutional Rights, 1870–1970," in George Thomas, John Meyer, Francisco Ramirez, and John Boli, *Institutional Structure: Constituting State, Society, and the Individual* (Beverly Hills, Calif.: Sage, 1987), 133–49.

67. See Miller, *Jefferson and Nature,* 172–73.

68. Jamie Goodwin-White, "Where the Maps Are Not Yet Finished: A Continuing

American Journey," in *The Immigration Reader: America in a Multidisciplinary Perspective,* ed. David Jacobson (Oxford: Blackwell Publishers, 1998), 415–29.

69. Miller, *Jefferson and Nature,* 172–75; Jefferson copied the quote into his commonplace book from an anonymously published *Historical Essay on the English Constitution.*

70. Jefferson also wrote,

> Were our State a pure democracy, in which all its inhabitants should meet together to transact all their business, there would yet be excluded from their deliberations, 1. Infants, until arrived at years of discretion. 2.Women, who, to prevent depravation of morals and ambiguity of issue, could not mix promiscuously in the public meetings of men. 3. Slaves, from whom the unfortunate state of things with us takes away the rights of will and of property. Those then who have no will could be permitted to exercise none in the popular assembly; and of course, could delegate none to an agent in a representative assembly. The business, in the first case, would be done by qualified citizens only.

"Thomas Jefferson to Samuel Kercheval, 1816," in Andrew A. Lipscomb and Albert Ellery Bergh, eds., *The Writings of Thomas Jefferson,* 20 vols. (Washington, D.C.: Issued under the auspices of the Thomas Jefferson Memorial Association of the United States, 1903–4), 15:71. The quotation in the text is from "Thomas Jefferson to John Hampden Pleasants, 1824," 16:28.

71. Alexis de Tocqueville, *Democracy in America,* ed. Philips Bradley (1945; New York: Vintage Books, 1990), 1:60; emphasis added.

72. Ibid., 61–63, 66–68.

73. See Eugene D. Genovese, *The Southern Tradition: The Achievement and Limitations of an American Conservatism* (Cambridge: Harvard University Press, 1994), 70; Genovese quotes from the speech "Speech on Surveys for Roads and Canals," January 30, 1824, in Russell Kirk, *John Randolph of Roanoke: A Study in American Politics,* 3d ed. (Indianapolis: Liberty Fund, 1978), 433.

74. See, generally, Frederick Merk, *Manifest Destiny and Mission in American History* (New York: Knopf, 1970).

75. Denis P. Duffey, "The Northwest Ordinance as a Constitutional Document," *Columbia Law Review* 95 (May 1995): 934–35.

76. The Homestead Act, signed into law by Lincoln in 1862, provided 160 acres of public land, essentially free, to independent settlers who had cultivated the land they had lived on for at least five years. Some six hundred thousand homestead farmers had claimed about 80 million acres of land by the end of the nineteenth century. In addition to age or family status qualifications, the settlers had to be citizens (or have applied for citizenship) to benefit under the Homestead Act.

77. "Ordinance of 1787: The Northwest Territorial Government," United States Code Annotated [USCA], passed by the Confederate Congress, July 13, 1787.

78. See Duffey, "Northwest Ordinance as a Constitutional Document," 929–67, and USCA, "Ordinance of 1787."

79. One can argue that intellectual history is of political and sociological interest only to the extent that ideologies, myths, and beliefs are embodied in institutions that shape political and social practices. Looking at how the landscape is physically shaped to accord with the myths, beliefs, and symbols that constitute a people is a particularly vivid form of evidence of how ideas become "embodied" and are "materialized" in social and political practices.

80. Peter S. Onuf, *Statehood and Union: A History of the Northwest Ordinance* (Bloomington: Indiana University Press, 1987), 49–51, 73–74.

81. Ibid., 4–5, 57–58, 77–80, 109–32; see also Deudney, "Philadelphian System."

82. See Onuf, *Statehood and Union,* on surveying and the "grid," 38–42, 53–54, 89–91, and on the concern with demarcating the land in order to nurture orderly and public-minded citizens, 28–33.

83. See Frederick Jackson Turner, *The Significance of the Frontier in American History* (New York: Frederick Ungar Publishing, 1963). The quoted text is from p. 51.

84. As one commentator on Turner's thesis put it, "The American was to control nature and yet be receptive to its occult power." See Harold P. Simonson, introduction to Turner, *Significance of the Frontier,* 12.

85. Turner, *Significance of the Frontier,* 45–46.

86. Ibid., 45–48.

87. These data are compiled from a count of the relevant statutes in the *Statutes at Large* (Washington, D.C.: Government Printing Office, 1789–1859). I am grateful to Zeynep Kilic for her assistance in collecting this information.

88. See discussion and quotations in Merk, *Manifest Destiny and Mission in American History,* 211–12; emphasis in original.

89. See generally James C. Scott, *Seeing Like a State: How Certain Schemes to Improve the Human Condition Have Failed* (New Haven: Yale University Press, 1998).

90. Ibid., 30–36.

91. Quoted in Elliot West, "American Frontier," in *The Oxford History of the American West,* ed. Clyde A. Milner II, Carol A. O'Connor, and Martha A. Sandweiss (New York: Oxford University Press, 1994), 145–46.

THREE NATURE'S NATION

1. Merrill D. Peterson, *The Jefferson Image in the American Mind* (New York: Oxford University Press, 1962), 168–74.

2. Garry Wills, *Lincoln at Gettysburg: The Words That Remade America* (New York: Simon and Schuster, 1992), 128–29.

3. Benedict Anderson's *Imagined Communities: Reflections on the Origins and Spread*

of Nationalism (London: Verso, 1983) has been a much celebrated book, a central part of which is an argument about the importance of newspapers in mediating and fostering nationalism. The media have been a technical means in engaging the public—as well as, we shall see, visual imagery such as paintings—but the *organic* basis of the imagination was the land, mediated through the direct and vicarious experience of nature; national monuments and parks, as discussed below, played a central part in that regard.

4. Abraham Lincoln, "Speech at Chicago, Illinois, July 10, 1858," in *Abraham Lincoln: Speeches and Writings,* ed. Don E. Fehrenbacher (New York: Literary Classics of the United States, 1989), 1:457.

5. Wills, *Lincoln at Gettysburg,* 66.

6. See ibid., 63–89, and John Brinckerhoff Jackson, *Landscape in Sight: Looking at America* (New Haven: Yale University Press, 1997), 163–74.

7. Barbara Novak, *Nature and Culture: American Landscape Painting, 1825–1875,* rev. ed. (New York: Oxford University Press, 1995).

8. Angela Miller, *The Empire of the Eye: Landscape Representation and American Cultural Politics, 1825–1875* (Ithaca, N.Y.: Cornell University Press, 1993). The 1839 traveler was an English captain, Frederick Marryat, author of *Diary in America,* ed. Jules Zanger (Bloomington: Indiana University Press, 1960); quoted in Miller (7).

9. J. Gray Sweeney, "The Nude of the Landscape Painting: Emblematic Personification in the Art of the Hudson River School," *Smithsonian Studies in American Art* 3, no. 4 (1989): 42–65.

10. Quoted and discussed in Miller, *Empire of the Eye,* 8–9, 286; emphasis added.

11. Sweeney, "Nude of the Landscape Painting ," 43, 59.

12. Novak, *Nature and Culture,* 15–17.

13. Quoted in Sweeney, "Nude of the Landscape Painting," 47, 49.

14. Miller, *Empire of the Eye,* 234–44, 286–87.

15. Novak, *Nature and Culture,* 19

16. Miller, *Empire of the Eye,* 3–12

17. Quoted in James M. Mayo, *War Memorials as Political Landscape* (New York: Praeger, 1988), 216.

18. This is evident in that Lincoln countenanced the continuation of slavery, in the short term, to aid the Union, by co-opting slave states without demanding immediate emancipation (unlike slaves states defeated in the field of battle). He even suggested, in the case of cooperating slave states, that "in [his] judgment, gradual, and not sudden emancipation is better for all." See his "Message to Congress, March 6, 1862," in *Speeches and Writings,* 2:307–8.

19. Abraham Lincoln, "First Inaugural Address, March 4, 1861," in ibid., 2:224; emphasis added.

20. Lincoln used the terms *chords* and *cords* indiscriminately; discussed in Wills, *Lincoln at Gettysburg,* 159.

21. Abraham Lincoln, "Annual Message to Congress, December 1, 1862," in *Speeches and Writings,* 2:406; emphasis added.

22. See Eugene D. Genovese, *The Southern Tradition: The Achievement and Limitations of an American Conservatism* (Cambridge: Harvard University Press, 1994), on the southern opposition to the "rationalizing" tendencies of the North. For a richly detailed description of one "plantation neighborhood," see James Everett Kibler, *Our Fathers' Fields: A Southern Story* (Columbia: University of South Carolina Press, 1998), chap. 11.

23. Some historians have questioned whether the Civil War was significant in creating a more egalitarian society, pointing to Jim Crow laws and other deprivations for black Americans in the decades following the war. However, as James McPherson points out, such critics are looking in the wrong end of the telescope, looking at the point of "arrival" of blacks (say in the early twentieth century) rather than at the baseline at the end of the Civil War. McPherson discusses this issue and the changes in the material and political status of blacks in his *Abraham Lincoln and the Second American Revolution* (New York: Oxford University Press, 1990), 16–22.

24. Abraham Lincoln, "Message to Congress in Special Session, July 4, 1861," in *Speeches and Writings,* 2:259.

25. Lincoln, "Speech at Chicago," in ibid., 1:581.

26. Abraham Lincoln, "Speech at Edwardsville, Illinois, September 11, 1858," in ibid., 1:457; emphasis added.

27. Abraham Lincoln, "Eulogy on Henry Clay at Springfield, Illinois, July 6, 1852," in ibid., 1:271.

28. Regarding the continuing play of the farmer and the homestead: it was Lincoln, after all, who signed into law the Homestead Act of 1862.

29. Slightly different versions of the Gettysburg Address are in circulation; the one quoted here is the mostly commonly cited. See discussion in Wills, *Lincoln at Gettysburg,* 191–203.

30. Quoted in Robert Pinsky, "Poetry and American Memory," *Atlantic Monthly* 284, no. 4 (1999): 63.

31. Wills, *Lincoln at Gettysburg,* 53–55

32. Quoted in McPherson, *Abraham Lincoln and the Second American Revolution,* 3.

33. Abraham Lincoln, "To Mrs. Lydia Bixby, November 21, 1864," in *Speeches and Writings,* 2:644.

34. Brander Matthews, *Poems of American Patriotism* (New York: Scribner's, 1900), 146.

35. See Norbert Elias, *The Civilizing Process: Sociogenetic and Psychogenetic Investigations,* rev. ed. (Oxford: Blackwell Publishers, 2000).

36. James Kettner, *The Development of American Citizenship, 1608–1870* (Chapel Hill: University of North Carolina Press, 1978).

37. George Mosse, *Fallen Soldiers: Reshaping the Memory of the World Wars* (New

York: Oxford University Press, 1990), chap. 3. Mosse notes that in Europe the first military cemetery was established only with the Franco-Prussian War in 1870–71, by which time there were already seventy-three military cemeteries in the United States. The cemeteries of the Franco-Prussian War were, however, almost an accident in that soldiers were buried where they fell. However, that cemetery was not a site for national commemoration. Only in World War I was an attempt made to bury the war dead systematically and commemoratively in military cemeteries, when on December 29, 1915, France mandated a perpetual resting place for the war dead. Other European countries soon followed in kind. This was more than fifty years after the United States had undertaken such an endeavor (45–46).

38. It is in this period, after the Civil War, that Americans developed a strong interest in celebrating the past and in developing what scholars term a "historical memory." This interest in history is presumably related to the evolution of a "mediated" nationhood in that symbols of nationhood—in this case memorabilia, historical writings, and the like—were sought out. See Michael Kammen, *Mystic Chords of Memory: The Transformation of Tradition in American Culture* (New York: Knopf, 1991), especially pt. 2.

39. I take the term "mind's I" (but not necessarily the authors' compositions of the term) from Douglas Hofstadter and Daniel C. Dennett, eds., *The Mind's I: Fantasies and Reflections on Self and Soul* (New York: Basic Books, 1981).

40. See John F. Sears, *Sacred Places: American Tourist Attractions in the Nineteenth Century* (New York: Oxford University Press, 1989), 2–11.

41. Jackson, *Landscape in Sight*,164.

42. Ibid., 165.

43. See David E. Stannard, *The Puritan Way of Death: A Study in Religion, Culture, and Social Change* (New York: Oxford University Press, 1977).

44. Jackson describes Hillhouse in *Landscape in Sight*, 166–68.

45. Quoted in ibid., 169.

46. Speech of Everett appears in full in Wills, *Lincoln at Gettysburg*, 213–47; emphasis added.

47. Ibid., 72–73. The graveyard set in nature, conjuring up a sense of the liminal state between life and death, nurtured an almost pleasing sadness that is evident in Everett's words, noted in the previous paragraph, that "you feel, though the occasion is mournful, that it is good to be here." It is a feeling that Søren Kierkegaard describes, in *Either/Or, Part I*, ed. and trans. Howard V. Hong and Edna H. Hong (Princeton: Princeton University Press, 1987), 77, in a different context, as that "ambiguity that is the sweetness in melancholy" (which has also been translated from Danish as "sweetness *of* melancholy"). This explains, in part, the desire to prolong national mourning.

48. Johan Huizinga, *The Waning of the Middle Ages* (London: Penguin Books, 1982), 215.

49. The reference here is to the biblical Aaron, the brother of Moses, who founded and led the Jewish priesthood (and thus "stood between the living and the dead").

50. Jackson, *Landscape in Sight,* 178.

51. See extracts of writings from the period and commentary in John W. Reps, *Washington on View: The Nation's Capital since 1790* (Chapel Hill: University of North Carolina Press, 1991), 16, 22; emphasis added.

52. Jackson, *Landscape in Sight,* 175–77.

53. See Withold Rybczynski, *A Clearing in the Distance: Frederick Law Olmsted and America in the Nineteenth Century* (New York: Scribner's, 1999).

54. Jackson, *Landscape in Sight,*180.

55. Mayo, *War Memorials as Political Landscape,* 80–89.

56. Kenneth E. Foote, *Sacred Ground: America's Landscapes of Violence and Tragedy* (Austin: University of Texas Press, 1997), 284–85.

57. The martial motif of monuments and public squares (and also the naming of streets after war heroes and battles) represents civic obligation; but it also represents independence of the Republic from outside forces. Civic politics is, in this regard, about place, or claims about place, and where "we" belong as individuals and as nations in the larger world. "Place and belonging" in this context refer to essentially the same phenomenon: place is about belonging, and belonging is about place.

FOUR SPATIAL RHYTHMS

1. Johan Huizinga, *The Waning of the Middle Ages* (London: Penguin Books, 1982), 136, 159.

2. Sanford Levinson, *Written in Stone: Public Monuments in Changing Societies* (Durham, N.C.: Duke University Press, 1998), 7.

3. Friedrich Nietzsche, *Untimely Meditations,* trans. R. J. Hollingdale (Cambridge: Cambridge University Press, 1983), 68.

4. Alfred Runte, *American Parks: The National Experience* (Lincoln: University of Nebraska Press, 1979), 24.

5. The number of parks and sites is approximate for reasons of jurisdiction and interpretation. The primary source for drawing on such figures is the National Park Service (NPS), but not all parks are under the jurisdiction of the NPS. The United States Holocaust Memorial Museum for example, discussed below, is an independent federal agency and does not have a direct relationship with the NPS; yet it is a major commemorative site and one associated with war. The American Battle Monuments Commission (ABMC) is a small independent federal agency responsible for 24 cemeteries of American war dead and 27 memorials, monuments, or markers, almost entirely on foreign soil. The ABMC's figures are not included by the NPS; nonetheless the sites are of a military nature, and some, such as the American cemetery and memorial over-

looking Omaha beach in Normandy, loom large in the American memory. Then there is the question of interpretation: should Lincoln's birthplace, a national historic site, be considered "military," in that he was associated so closely with the Civil War? (The NPS does indeed include it as a "Civil War" site.) The figure of approximately 70 parks noted here is decidedly on the conservative side and tends to understate the military theme in federal and public commemoration—it includes revolutionary and Civil War sites as defined by the NPS and other sites such as the Korean war memorial, the Vietnam War monument, and the National World War II Memorial. It does not include the ABMC cemeteries and monuments, for example. Similarly, the figure of approximately 400 parks, monuments, and sites refers to NPS-controlled entities (the NPS in its *Index,* published in 1999, cited below, lists 378 "areas" under its control; the NPS Web site by the end of the year 2000 listed 424 areas, including parks, monuments, trails, sites, and the like, under its jurisdiction). In sum, the figures here are conservative and, if anything, understate the trends and characteristics described here, especially the prominence of military and war-related markers of commemoration (not only their numbers but their centrality in the nation's symbolic heart, Washington, D.C.). See U.S. Department of the Interior, *The National Parks: Index, 1999–2001* (Washington, D.C.: Government Printing Office, 1999), and the NPS Web site: www.nps.gov.

6. www.nps.gov. A good number of parks overlap different categories. Manzanar National Historic Site, for example, is categorized both as a human rights park and as a part of the Asian American heritage.

7. Twenty parks in the United States are designated by the United Nations as World Heritage Sites.

8. Albert Boime, *The Unveiling of the National Icons: A Plea for Patriotic Iconoclasm in a Nationalist Era* (New York: Cambridge University Press, 1998), 1–24.

9. See discussion in David Jacobson, "Protestantism, Authoritarianism, and Democracy: A Comparison of the Netherlands, the United States, and South Africa," *Religion* 17 (July 1987): 275–301.

10. Aldous Huxley, *The Perennial Philosophy* (New York: Harper Colophon Books, 1970), 21; or, one add the words of Meister Eckhart, the medieval mystic and theologian, "The Ground of God and the Ground of the Soul are one and the same" (Huxley, 12).

11. James Hutchison Stirling, *Secret of Hegel: Being the Hegelian System in Origin, Principle, Form, and Matter* (London: Longman, Green, Longman, Roberts, & Green, 1865), vol. 1, sec. 2, chap. 1, 345; emphasis in the original. Contrast this statement of Stirling, in his analysis of Hegel, with the following: "*Being* is the indefinite *Immediate;* it is devoid of definiteness" (319); "Being, too, can be thrown into the all from which abstraction has been already made; then remains Nothing," and "Being is determined as Infinite Being" (346); and "Being . . . is only *indefinite immediacy,* as Nothing in the same way is *immediate indefiniteness,*" and, consequently, "Being and Nothing are thus the

same" (vol. 2, sec. 3, chap. 1, 77; all emphases in original). In this context one could infer, then, that *the Ground* gives Being a presence, no longer devoid of definiteness.

12. Joseph Tilden Rhea, *Race Pride and the American Identitiy* (Cambridge: Harvard University Press, 1997), 8–9, 20. The quotation from the NPS brochure appears in Rhea; he cites "Custer Battlefield National Cemetery" (Washington D.C.: National Park Service, [ca. 1945]), 1.

13. The phrase "rational restlessness" is used in a different context—on religion and the development of capitalism—in R. H. Tawney, *Religion and the Rise of Capitalism: A Historical Study* (New York: Harcourt, Brace, 1952).

14. See Edward Tabor Linenthal, *Sacred Ground: Americans and Their Battlefields* (Urbana: University of Illinois Press, 1991), 150–53.

15. Linenthal discusses and quotes (drawn on here) from the NPS "master plans" in ibid., 153–54.

16. The term is from Levinson, *Written in Stone,* 78.

17. Ibid.; see also Linenthal, *Sacred Ground,* 142.

18. Rhea, *Race Pride,* 8–18.

19. *Little Bighorn National Monument,* Public Law 102–201, 102d Cong., 1st sess., 1991.

20. Levinson, *Written in Stone,* 10–12.

21. "Tolerance" in this context entails very real restraints. These restraints are elaborated on in the next chapter.

22. Arthur Amiotte, "Preamble," Little Bighorn Battlefield Indian Memorial Advisory Committee, Design Competition Packet, National Park Service, United States Department of Interior, Washington, D.C., 1996; emphasis added.

23. The very sense among some today that indigenous and nomadic people have a more "natural" relationship to the land than territorial states may reflect the more fluid character of place and belonging in the contemporary world, an issue we return to in Chapter 5.

24. Little Bighorn Battlefield Indian Memorial Design Competition, "Jury Report," Billings, Montana, February 14, 1997 (including design statement submitted by winning entry); and U.S. Department of Interior, "News Release," National Park Service, Little Bighorn Battlefield National Monument, February 17, 1997.

25. Little Bighorn Battlefield National Monument Advisory Committee, "Program and Announcement of Indian Memorial Winners: Design Discussion," March 22, 1997, Denver, Colorado; recorded by Bruce Chandler, Institute of American Indian Studies, University of South Dakota.

26. Ibid.

27. Amiotte, "Preamble."

28. Violence must be viewed here, sociologically speaking, not simply as an instrument, a means to some other purpose or goal; it is, in a sense, an expression, in its most

extreme form, of the nation itself. Perhaps one could go so far as to say that it is a "worldly mysticism." It is mysticism in its destruction and transcendence of bounded categories of everyday life; it is worldly in its political and collective orientation. However, this sense of violence is generally in the abstract; in most cases the actual experience of violence, notably in World War I, is a sobering moment.

29. See Norbert Elias, *The Civilizing Process: Sociogenetic and Psychogenetic Investigations,* rev. ed. (Oxford: Blackwell Publishers, 2000).

30. See Jan C. Scruggs and Joel L. Swerdlow, *To Heal a Nation: The Vietnam Veterans Memorial* (New York: Harper Collins, 1985).

31. See www.nps.gov.

32. Interestingly, pacifism and antiviolence grew out of otherworldly, mystical, highly individualistic Protestant sects such as the Quakers, who stressed the subjective experience of the Spirit, in contrast to the worldly and civic-minded Calvinist churches, who believed in preparing for combat against what were for them evident forces of darkness (such as the Catholic Church).

33. Maya Lin, *Boundaries* (New York: Simon and Schuster, 2000), chap 4, p. 5.

34. See John Ayto, *Dictionary of Word Origins* (New York: Little, Brown, 1990), 412–13.

35. Quoted in Thomas B. Allen, *Offerings at the Wall: Artifacts from the Vietnam Veterans Memorial Collection* (Atlanta: Turner Publishing, 1995), 9. Shakespeare wrote in *The Tempest,* to quote in full, "In the dark backward and abysm of time."

36. Ibid, 7.

37. See www.nps.gov/mrc/vvmc/vvmc.htm (posted in 2000).

38. See the *Oxford English Dictionary* under "pilgrim" and "pilgrimage."

39. The most commonly left items are listed at www.nps.gov/mrc/vvmc/faqs.htm (posted in 2000).

40. Mircea Eliade, *The Sacred and the Profane: The Nature of Religion* (New York: Harcourt, Brace and World, 1957), 57.

41. Maya Lin, "Making the Memorial," *New York Review of Books,* November 2, 2000, 33, 35.

42. See discussion in Scruggs and Swerdlow, *To Heal a Nation,* 80–84.

43. Maya Lin's statement that she wanted "to create spaces for people to think without telling them what to think" is telling in this regard. This statement is an epigraph to a Public Broadcasting System (PBS) Web site designed to generate, through contributions by readers, Vietnam War–related stories—a popular history of sorts which eschews a national narrative; see www.pbs.org/pov/stories/index.html (posted in 2000).

44. Johan Huizinga, *Homo Ludens: A Study of the Play Element in Culture* (Boston: Beacon Press, 1955), 15.

45. Quoted in Elizabeth Hutton Turner, *Georgia O'Keeffe: The Poetry of Things* (New Haven: Yale University Press, 1999), 1. One could infer from the sentiment expressed in O'-

Keeffe's statement and others like it two (related) points: first, simplicity of form (which is abstraction) captures the essence, the soul, as abstraction is more "formless" than complexity—thus most suggestive of the All in All. Second, by simplicity of form we evoke, paradoxically, the "spiritual" or ethereal qualities of matter. On a different but related plane Kierkegaard, in "The Intermediate Erotic States or the Musical Erotic," notes that "the more abstract an idea is, the less the probability. But how does the idea become concrete? By being permeated by the historical. The more concrete the idea, the greater the probability." Abstraction, of course, can express different states aside from loss; Kierkegaard suggests that sensuality expresses the zenith of abstraction. "The most abstract idea conceivable is the sensuous in its elemental originality," and, he notes elsewhere, "to see her and to love her are the same; this is the moment. In the same moment everything is over, and the same thing repeats itself indefinitely . . . sensuous love is disappearance in time." Søren Kierkegaard, *Either/Or, Part I,* edited and translated by Howard V. Hong and Edna H. Hong (Princeton: Princeton University Press, 1987), 55, 56, 95.

46. The quote is from the Vietnam Women's Memorial Project, Inc., "Vietnam Women's Memorial Project: A Legacy of Healing and Hope," Washington, D.C., 2000, formerly at www.albany.net/~deavila/viet.html.

47. Deborah K. Dietsch, "Memorial Mania," *Architecture,* September 1997, 94–97.

48. Charles Fishman, ed., *Blood to Remember: American Poets on the Holocaust* (Lubbock: Texas Tech University Press, 1991), vii.

49. Owen Dodson, "Jonathan's Song," in his *Powerful Long Ladder* (New York: Farrar, Straus and Co., 1946), 89–90.

50. Edward T. Linenthal, *Preserving Memory: The Struggle to Create America's Holocaust Museum* (New York: Viking, 1995), 1.

51. Ibid.

52. See discussion, and quotations from the debate, in ibid., 57–72.

53. See United States Holocaust Council, *A National Commitment to Remembrance: Official Groundbreaking of the United States Holocaust Memorial Museum* (Washington, D.C.: United States Holocaust Memorial Council, 1986).

54. States are *especially* subject to principles of human rights; ironically, it is substate authorities that can more easily discriminate, such as the province of Quebec favoring francophone culture. Quebec, were it to become an internationally recognized "sovereign" state, would likely face opprobrium on the international stage for such decidedly slanted policies.

55. Linenthal, *Preserving Memory,* 67, 171–82.

56. Ibid., 15, 36, 88–89, 199–205; quotation, cited by Linenthal, is from Richard Rubenstein, *The Cunning of History: The Holocaust and the American Future* (New York: Harper and Row, 1975), 6.

57. Freed is quoted in James E. Young, *The Texture of Meaning: Holocaust Memorials and Meaning* (New Haven: Yale University Press, 1993), 283.

58. See the United States Holocaust Council, *National Commitment to Remembrance,* 2.

59. It is in this frame that, in the words of Owen Dodson, quoted previously, "Jew is not a race / Any longer—but a condition" (see n. 49 above).

60. Roy and Alice Eckardt, "Travail of a Presidential Commission Confronting the Enigma of the Holocaust," *Encounter* 42, no. 2 (1981), 103–14.

61. See G. K. Woods, comp., *Personal Impressions of the Grand Cañon of the Colorado River Near Flagstaff, Arizona As Seen Through Nearly Two Thousand Eyes, and Written in the Private Visitors' Book of the World-Famous Guide Capt. John Hance* (San Francisco: Whitaker & Ray Co., 1899), 43–95.

62. Stephen J. Pyne, *How the Canyon Became Grand* (New York: Penguin Books, 1998), xiv, 84, 144–58; on the "unique privilege" of being American, Pyne quotes Joseph Wood Krutch, *Grand Canyon: Today and All Its Yesterdays* (New York: William Morrow & Co., 1957), 275.

63. National Park Service, Division of Interpretive Planning, Harpers Ferry Center, Harpers Ferry, West Virginia, "A Plan for the Interpretation of Grand Canyon National Park," 1996, 3 (emphasis added).

64. See, for example, National Park Service, Grand Canyon National Park, "Annual Statement of Interpretation and Visitor Services," 1986, p.33.

65. Ellen Sealey, interpretive planner; Sara Stebbins, research librarian; and Stewart Fritts, interpretive specialist, all of the Grand Canyon National Park, National Park Service, interview with author, February 2000.

66. Margaret Littlejohn and Gary Machlis, "A Diversity of Visitors: A Report on Visitors to the National Park System," National Park Service, Moscow, Idaho, 1990.

67. National Park Service, Grand Canyon National Park, "Comprehensive Interpretive Plan: A Foundation for Interpretive Programming and Planning," January 2000.

68. Clive Irving, "The Century of Travel," *Condé Nast Traveler* (December 1999): 123–37.

FIVE INTANGIBLE PROPERTY

1. Michael Pollan, "The Triumph of Burbopolis," *New York Times Magazine,* April 9, 2000, 55. Philip Langdon, *A Better Place to Live: Reshaping the American Suburb* (New York: Harper Collins, 1994), states that by the 1990s three-quarters of the U.S. population lived in metropolitan areas, and of those, two-thirds lived in the suburbs. The U.S. Census defines suburbs as metropolitan areas outside the city center.

2. See Richard John Neuhaus, *The Naked Public Square: Religion and Democracy in America* (Grand Rapids, Mich.: W. B. Eerdmans Publishing Co., 1984).

3. The quotation, used here in a completely different context, is from a fifth-cen-

tury Chinese art critic, Hsieh Ho, and is cited in Elizabeth Hutton Turner, *Georgia O'-Keefe: The Poetry of Things* (New Haven: Yale University Press, 1999), 90.

4. Pollan, "Triumph of Burbopolis," 55.

5. Sean-Shong Hwang and Steve M. Murdock, "Racial Attraction or Racial Avoidance in American Suburbs," *Social Forces* 77, no. 2 (December 1998): 541–66.

6. Richard D. Alba, John R. Logan, Brian J. Stults, Gilbert Marzan, and Wenquan Zhang, "Immigrant Groups in the Suburbs: A Reexamination of Suburbanization and Spatial Assimilation," *American Sociological Review* 64 (June 1999): 446–60.

7. See J. G. A. Pocock, "Authority and Property: The Question of Liberal Orgins," in his *Virtue, Commerce, and History: Essays on Political Thought and History, Chiefly in the Eighteenth Century* (New York: Cambridge University Press, 1985), 51–71.

8. WIPO, "Millennium Message from Dr. Kamil Idris, Director General of WIPO," at www.wipo.int/eng/dg_idris.htm (posted in 2000).

9. Jamie Goodwin-White, "Where the Maps Are Not Yet Finished: A Continuing American Journey," in *The Immigration Reader: America in a Multidisciplinary Perspective*, ed. David Jacobson (Oxford: Blackwell Publishers, 1998), 415–29. As Steven Grosby pointed out (personal communication), citizenship is about "temporal belonging" as well, in that traditionally the nation's past became, at least in principle, the past of the immigrant.

10. It is also the powerful "identifying" aspect of citizenship (identity and its documentation being a central preoccupation of the modern state) which is interwoven with the territorial and political axes of the nation, which made the issue of incorporation of outsiders so critical. See John Torpey, *The Invention of the Passport: Surveillance, Citizenship, and the State* (New York: Cambridge University Press, 1999).

11. Charles Tilly, *Durable Inequality* (Berkeley: University of California Press, 1998).

12. The following statistical snapshots indicate the scope of these developments. In a relatively recent period, particularly in the 1990s, countries that account for the bulk of foreign populations in the United States (and in Western Europe)—such as Mexico, Turkey, Italy, the Dominican Republic, Russia, El Salvador, and Columbia—have legalized dual citizenship or nationality; and, as is widely reported, foreign and immigrant populations have changed the ethnic hues and social landscapes in the United States especially since 1965. In the United States, according to Census Bureau figures, the foreign-born population was 90 percent white in 1970, compared with 50 percent in 1990. The Hispanic share among the foreign-born jumped from 15 percent in 1970 to 40 percent in 1990. Asians account for another 25 percent of the foreign-born in the United States today.

13. See Rey Koslowski, *Migrants and Citizens: Demographic Change in the European State System* (Ithaca, N.Y.: Cornell University Press, 2000), chap. 7.

14. Quoted in the *New York Times*, April 14, 1998, A12.

15. See Arjun Appadurai, *Modernity at Large: Cultural Dimensions of Globalization* (Minneapolis: University of Minnesota Press, 1996).

16. This paragraph and the following five paragraphs draw partly from David Jacobson, "New Border Customs: Migration and the Changing Role of the State," *UCLA Journal of International Law and Foreign Affairs* 3 (Fall/Winter 1998–99): 443–58. Administrative and judicial rules grow in mediating the kaleidoscope-like complexity of a social world with an almost geometric increase in the number of actors and disparate social, economic, and political forces. This parallels the evolution of domestic law within nation-states since at least the nineteenth century, when legal mechanisms of control became tighter and denser with growing economic and social differentiation, specialization, and complexity.

17. Emile Durkheim, *The Division of Labor in Society* , trans. W. D. Halls (1893; New York: Free Press,1984), 161.

18. The same applies to international treaties on human rights: states may have "contracted" into human rights when those treaties were explicitly state-centric, but with changing circumstances individuals have increasingly been able to make claims on states, thus contributing to changing the identity of the states themselves. Yet states have not—not in a single case, at least in the West—withdrawn their participation in such human rights regimes. The state has become enmeshed, unwittingly, in a set of institutional arrangements that define what it means to be a (legitimate) state.

19. See Eric Hobsbawm and Terence Ranger, eds., *The Invention of Tradition* (New York : Cambridge University Press, 1992), and John W. Meyer, John Boli, George M. Thomas, and Francisco O. Ramirez, "World Society and the Nation-State," *American Journal of Sociology* 103 (July 1997): 144–81.

20. This is a phenomenon in Europe as well, especially under the impact of regional courts. See the discussions in Anne-Marie Slaughter, Alec Stone Sweet, and J. H. H. Weiler, *The European Court and National Courts—Doctrine and Jurisprudence: Legal Change and Its Social Context* (Oxford: Hart Publishing, 1998). On the broader global phenomenon of the legal expansion, see Jacobson, "New Border Customs."

21. See U.S. Census Bureau, *Statistical Abstract of the United States: 1999* (Washington, D.C.: Government Printing Office, 1999), 227, for the caseload figures from 1995, and U.S. Census Bureau, *Statistical Abstract of the United States: 1990* (Washington, D.C.: Government Printing Office, 1990), 183, for the caseload figures from 1970.

22. The figures are from a Westlaw database search by the author.

23. David Vogel, "The 'New' Social Regulation in Historical and Comparative Perspective," in *Regulation in Perspective,* ed. Thomas K. McCraw (Cambridge: Harvard University Press, 1981), 161–62; cited in Michael Schudson, *The Good Citizen: A History of American Civic Life* (Cambridge: Harvard University Press, 1998), 267.

24. See, in general, Lawrence M. Friedman, *Total Justice* (New York: Russell Sage Foundation, 1994).

25. Erving Goffman, *Relations in Public: Microstudies of the Social Order* (New York: Basic Books, 1971), 6.

26. See David Jacobson, "New Frontiers: Territory, Social Spaces, and the State," *Sociological Forum* 12, no. 1 (1997): 121–33.

27. Paul Colomy and J. David Brown, "Goffman and Interactional Citizenship," *Sociological Perspectives* 39, no. 3 (1996): 371–81.

28. See Hannah Arendt, "The Public and the Private Realm," in *The Human Condition* (Chicago: University of Chicago Press, 1957), 22–78.

29. See Thomas Franck, "Clan and Superclan: Loyalty, Identity, and Community in Law and Practice," *American Journal of International Law* 90 (July 1996): 359–77.

30. See Gérard Noiriel, *The French Melting Pot: Immigration, Citizenship, and National Identity,* trans. Geoffroy de Laforcade (Minneapolis: University of Minnesota Press, 1996).

31. This is a point worth pondering, considering academic claims these days about "discovering" past notions of "essentialism" and supposedly present-day insights about constructed identities.

32. Ian F. Haney López, *White by Law: The Legal Constructions of Race* (New York: New York University Presses, 1996), 1–3.

33. *Ex Parte Mohriez,* 54 F. Supp. 941.

34. See, for example, *United States v. Bhagat Singh Thind,* 261 U.S. 204 S.Ct. 338, which took place in 1923.

35. See, for example *U.S. v. Baecker,* 55 F. Supp. 403. This case, which took place in 1944, concerned German immigrants whose citizenship was revoked because of their activities in a Nazi organization, the German-American Bund. It was the "nature and extent" of their participation in the organization that the court pointed to, not the membership in the organization as such. The court concluded in this case that one of the defendants, Paul Gies, "at the time of making and filing his petition for naturalization [in 1929], was not attached to the principles of the Constitution of the United States and well disposed to the good order and happiness of the United States. It was not his intention to become a citizen of the United States and to renounce absolutely and forever all allegiance and fidelity to any foreign prince, potentate, state or sovereignty, and particularly to the German Reich, of which he at that time was a citizen, and it was not his intention to reside permanently in the United States." The court further noted that it "finds that defendant has at all times since he has been in the United States been attached to German National Socialistic ideas and principles, which are incompatible with attachment to the principles of the Constitution of the United States." The case illustrates both the political distinctiveness of the American landscape and the singular loyalties that that political identity demanded.

36. See Margaret Talbot, "Baghdad on the Plains," *New Republic,* August 11, 1997, 18–20.

37. See *State of Nebraska v. Latif Al-Hussaini,* 579 N.W. 2d 561.

38. Board of Immigration Appeals Decision, *In Re Fauziya Kasinga,* June 13, 1996; 35 I.L.M. 1145 (1996).

39. State hate crimes statutes and federal initiatives are listed on the Web site of the Anti-Defamation League, www.adl.org (posted 2001).

40. Jeff Spinner considers the role of the state and the formation of identity in a different context in *The Boundaries of Citizenship: Race, Ethnicity, and Nationality in the Liberal State* (Baltimore: Johns Hopkins University Press, 1994).

41. Joseph R. Levenson, *Confucian China and Its Modern Fate: A Trilogy* (Berkeley: University of California Press, 1968), 1:xiii-xix.

42. See, generally, Tilly, *Durable Inequality,* and Mark Granovetter, "Economic Action and Social Structure: The Problem of Embeddedness," *American Journal of Sociology* 91 (1985): 481–510.

43. See, in general, Georg Simmel, *The Philosophy of Money,* trans. Tom Bottomore and David Frisby (London: Routledge, 1990).

44. Case in point: from World War I there is a steady accumulation in the number of multilateral international legal treaties that are deemed "significant," and a dramatic increase after World War II. Suggesting an increasingly dense global legal environment, the number of "significant" multilateral treaties in force rose from 187 in 1950 to almost 800 by 1988. The uncompiled listings are in *M. J. Bowman and D. J. Harris, Multilateral Treaties: Index and Current Status* (9th Cum. Supp. 1993). I am grateful to David John Frank for making this information available to me. See also the discussion in Jacobson, "New Border Customs."

45. See Robert D. Putnam, *Bowling Alone: The Collapse and Revival of American Community* (New York: Simon and Schuster, 2000).

46. Michael Schudson argues that rather than civic decline, the "web" of citizenship has expanded through this process. In some sense we have a semantic argument here; citizenship has been transformed at the very least. Republican politics has declined, and the citizenship associated with that model has thus declined as well. See Schudson, *Good Citizen,* 240–93.

47. I draw this formulation of agency from James E. Block, *A Nation of Agents: The American Path to a Modern Self and Society* (Cambridge: Harvard University Press, forthcoming). Block wants to recast the American historical project from its beginnings as driven by agency rather than liberty. Regardless, in most of American history it certainly can be argued that even if this interpretation is correct—and it is an intriguing one—agency could be realized in a republican framework. Agency *was* collective; it was expressed through *national* self-determination. Now, it is argued here, agency has been disengaged from the republican model.

48. Friedman, *Total Justice,* 22.

49. David Popnoe, *Private Pleasure, Public Plight* (New Brunswick, N.J.: Transaction

Books, 1985), 118; Popnoe takes a decidedly negative view of this "privatization," like most intellectuals, noting how "privatization is a dominant motif" and that "streets and sidewalks that once provided public pedestrian interaction and even entertainment have for the most part been abandoned to the utter privacy of the automobile; public parks have fallen into disuse and misuse; town squares have become the mostly ornamental appendages of commercialism" (117).

50. Galya Benarieh Ruffer, "Virtual Citizenship: Migrants and the Constitutional Polity" (Ph.D. diss., University of Pennsylvania, 2001).

51. See the interview with Gerhard Casper in Jodi Wilgoren, "Exiting Stanford, Noting Many Challenges Ahead," *New York Times,* September 22, 1999.

52. Editorial, "The Allure of Place in a Mobile World," *New York Times,* December 15, 1999, A30.

53. I thank Carolyn Forbes for her suggestive comments and formulation of this paragraph.

54. See Philip Q. Yang, *Post-1965 Immigration to the United States: Structural Determinants* (Westport, Conn.: Praeger, 1995).

CODA THE LABYRINTH OF THE SOUL

1. Octavio Paz, *The Labyrinth of Solitude* (New York: Grove Press, 1985), 9.

2. This paradox is also expressed in a different form in a question only the uncluttered and distilled mind of a child could produce; my son asked me, when he was four or five years old, "What happens when you disappear in your own imagination?"

3. Antonio R. Damasio, *The Feeling of What Happens: Body and Emotion in the Making of Consciousness* (New York: Harcourt Brace & Co., 1999), 4–18.

4. See, generally, ibid.; quoted text on p. 17.

5. It is in our ability to place ourselves outside ourselves—to imagine different ways of being—which makes us human. We can "imagine" ourselves in the place of others; indeed, we have to in order to engage them. We have to talk the same "language"—to use the scripts and the symbols that will be mutually comprehensible. We extend ourselves into the drama of the play, of life. Thus the "extended consciousness" has a social or sociological as well as biological component. In a sense, in the concept of extended consciousness we have the biological basis of human forms of association.

6. It is a location that always has to be justified, and this brings to the fore the sociological significance of place.

7. William Blake, "Auguries of Innocence," in *William Blake: The Complete Poems,* ed. Alicia Ostriker (New York: Penguin Books, 1977), 506.

8. The term *suprahistorical* here is used differently than it is by Nietzsche. "Suprahistorical man," wrote Nietzsche, "sees no salvation in the process and for whom . . . the world is complete and reaches its finality at each and every moment." In this context he

quotes the Italian poet and philosopher Giacomo Leopardi, who stated, "Our being is pain and boredom and the world is dirt—nothing more." The use of the word *dirt* is telling; it is a desire to see the world as it *is*, stripped of mythology—or what is understood to be "is" and "mythology." See Friedrich Nietzsche, *Untimely Meditations*, trans. R. J. Hollingdale (Cambridge: Cambridge University Press, 1983), 66.

9. Henry David Thoreau, *Walden and Civil Disobedience*, ed. Owen Thomas (New York: W. W. Norton, 1966), 204.

10. Quoted in Monique Laurent, *Rodin*, trans. Emily Read (New York: Konecky and Konecky, 1989), 100.

11. Quoted in John E. Bowlt and Evgeniia Petrova, *Painting Revolution: Kadinsky, Malevich, and the Russian Avant Garde* (Bethesda, Md.: Foundation for International Arts and Education, 2000), 190.

12. Apparently no contemporaneous documented explanation exists as to the purpose of these church labyrinths.

INDEX